Stories of Human Resilience in a Changing Climate

Stories of Human Resilience in a Changing Climate offers a captivating glimpse into the human face of climate change.

Despite extreme climate-related events witnessed globally, many struggle to connect such global phenomena to their own lives in a tangible way. As a result, there is still a strong disconnect between the general awareness of climate change and the sense of urgency that inspires action. This book fills a critical void in the field by sharing 11 real-life experiences across five continents of the world, featuring ordinary people facing and adapting to the consequences of climate change in its different forms. Through gripping accounts of a nomad having to relocate from the desert to the nearest city, an indigenous person fighting back for land rights, and a winemaker finding creative solutions to increasing heatwaves, among other stories, the book humanises the challenges of climate change. By weaving scientific data with personal stories, the chapters deliver a compelling call to move beyond mere acknowledgment of the problem towards transformative action that respects the balance of ecosystems, fosters sustainable practices, and redefines our relationship with nature. Overall, the book highlights the urgent need for global awareness on what ultimately represents one of the most pressing challenges of our time, leaving readers inspired and empowered to create a more sustainable and resilient future.

This accessible and impactful book will be of great interest to students and scholars of climate change and environmental sciences and communication, as well as community leaders and NGOs wanting to understand practical examples and insights of climate change phenomena.

Views and opinions expressed are those of the authors only and do not necessarily reflect those of the European Union or the European Commission.

Silvia Anna Ainio is a climate change and sustainable finance expert currently working at the European Commission, Brussels.

Susanna Gionfra is a biodiversity and nature-based solutions expert currently working at the European Commission, Brussels.

Routledge Focus on Environment and Sustainability

Sustainable Marketing and the Circular Economy in Poland
Key Concepts and Strategies
Anita Proszowska, Ewa Prymon-Ryś, Anna Kondak, Aleksandra Wilk and Anna Dubel

Risk Management for Water Professionals
Technical, Psychological and Sociological Underpinnings
Anna Kosovac

Climate Perspectives from the Congo Basin
Bila-Isia Inogwabini

Everyday Agri-Environmental Governance
The Emergence of Sustainability through Assemblage Thinking
Jérémie Forney, Dana Bentia, and Angga Dwiartama

Electric Cars and the Resource Challenge
Theo Henckens

Regulating Foreign Direct Investment for Development
Perspectives from Bangladesh
Nakib Mohammad Nasrullah and Mia Mahmudur Rahim

True Cost Accounting
Assessing Environmental and Social Damages of Production
Amelie Michalke, Jennifer Kunz, and Tobias Gaugler

Stories of Human Resilience in a Changing Climate
Silvia Anna Ainio and Susanna Gionfra

For more information about this series, please visit: www.routledge.com/Routledge-Focus-on-Environment-and-Sustainability/book-series/RFES

Stories of Human Resilience in a Changing Climate

Silvia Anna Ainio and Susanna Gionfra

Routledge
Taylor & Francis Group
LONDON AND NEW YORK

First published 2026
by Routledge
4 Park Square, Milton Park, Abingdon, Oxon OX14 4RN

and by Routledge
605 Third Avenue, New York, NY 10158

Routledge is an imprint of the Taylor & Francis Group, an informa business

British Library Cataloguing-in-Publication Data
A catalogue record for this book is available from the British Library

ISBN: 9781032797519 (hbk)
ISBN: 9781032797526 (pbk)
ISBN: 9781003493662 (ebk)

DOI: 10.4324/9781003493662

Typeset in Times New Roman
by KnowledgeWorks Global Ltd.

Contents

Figures

Acknowledgements

First and foremost, we would like to express our deepest gratitude to Clara Luz for her invaluable assistance in revising and refining our research sources. Her contributions significantly enhanced the overall quality of this work.

We are also sincerely grateful to Maria Nuñez Rodrigues for her beautifully designed map, which provided a cohesive visual representation of the diverse stories narrated in this book.

This book would not have been possible without the many individuals who contributed to it. We are especially thankful to Brahim El Gasmi, whose story served as the initial inspiration for our journey to collect and share individual narratives of resilience from around the world.

We are also thankful to Ernesto Rollando, who helped connect us with Jonathan Karigu Gachau in Kenya; Eduarda Zoghbi, whose dedication to supporting Indigenous women in Brazil led us to Vanda Witoto; Michael Tanner, who worked tirelessly to help us reach Santiago Morales; Sarah Faber, who introduced us to Evelio Morales; Kruti Munot, whose enthusiasm for this project brought us to Vincy Abraham; Lisa Klinger, whose deep love for Mongolia connected us with Bekbolat Bugibay; Giulia Meloni, who supported us in revising the chapter where we interviewed Marilena Barbera; and Filimone Tuivanualevu, who travelled to Tukuraki to meet Vilmaina Ratu.

Above all, we are profoundly grateful to the remarkable individuals we interviewed for this book: the ones we mentioned above, as well as Zack Porter and Koji Kinjo.

A special thanks to Sebastian Enning for believing in this project from the very beginning – specifically, from the trip to Morocco that set everything in motion – and encouraging us to take it forward.

Finally, we would like to thank each other for the dedication, patience, and unwavering support that carried us through this journey – one built upon a foundation of true friendship.

Introduction

The history of our planet may appear to us as a single, linear, and uniform narrative. However, what we are dealing with is a very rich mosaic of stories – those of individuals and communities, landscapes and ecosystems, biology and geology, philosophy and politics. Personal experiences, in particular, offer invaluable insights into the communities people belong to, the environments they inhabit, and the culture and traditions they cherish, often conveying far more elements than just an engaging individual tale. Yet, only a small fraction of these stories make their way into books, while the vast majority remain untold.

Once we get to learn about these experiences, however, they never get erased from our brain: they move some inner chords, and the main message can last forever within us. This is how the story of Brahim, a nomad from the Saharan desert, sparked our interest in writing this book in the first place. We met him on a trip to Morocco, where he was our guide through a night in the Erg Chigaga dunes. Despite the day having started off on a bad note, when Brahim spoke to us about his own challenging experience in the desert, something started to change. Suddenly, his way of seeing life managed to sweep away all other thoughts, and the knowledge he had about life in the desert became for us very well-targeted and well-timed advice.

Once back from Morocco, we both started to question ourselves: how could it be that someone so distant from our everyday life had managed to leave such a profound impact on us? One possible explanation was that Brahim was opening up and telling us his own personal story – a story of resilience, but also of great hardship, which despite being so unique and to some extent distant, was somehow still relatable. It was a story about his struggles facing a more-than-ever changing climate, which was threatening the existence of the very essence of his people: the nomads. When listening to him, we could grasp how powerful this message was as it showcased a real-life experience potentially connected not only to us but also to the hearts of many others.

This is where the challenge of this book lies: transmitting and preserving the untold climate stories that shape our world. All the stories in this book feature ordinary people, which differ greatly from one another in lifestyles, age,

DOI: 10.4324/9781003493662-1

beliefs, culture, and values. Yet, all of them are opening up and telling experiences which start with change and end with resilience, reminding us how the impacts of climate change are felt in every corner of the world.

In an age where we may feel overwhelmed by a wide but often contradictory knowledge, this book starts from the perspective of a single human being, addressing his or her experience facing the often-dire consequences of a changing climate, to then provide critical information about the phenomena they are dealing with. This marks a crucial starting point for recognising the profound interconnectedness between nature and humanity. It encourages us to view the climate change crisis not as a distant, abstract issue, but as an immediate and tangible consequence of our exploitative approach to the natural world.

What connects all stories together and brings them closer to the readers is rooted in the power of storytelling. Stories are as old as human civilisation itself: they have shaped our common heritage and defined how our ancestors preserved their traditions and norms, regardless of how different these were. They are also part of our first approach to life, in the form of kids' fairy tales: the very first taste of the culture and history of the civilisation we are in. For this reason, the mosaic of stories included in this book holds an incredible power: creating empathy and moving those chords that can leave a lasting mark on readers' minds.

While the stories in this book send a message of urgency for dealing with climate change and adapting to it, they also offer a glimpse of hope by portraying the endless resilience of human beings. Whether it's about dedicating a lifetime to finding ingenious solutions to save a coral reef or simply accepting the need to adapt one's livelihood to the growing threats of floods, all individuals in this book are pragmatic, acknowledging how there is no going back to some nostalgic past. Yet, as such, they are committed to making the most of their present to build a resilient future.

Another common trait of the individuals presented in the following pages is their strong sense of belonging to a community. As they understand the vastness of the challenge, they acknowledge how this poses a threat to the collectivity they live in – and not only to the single individual – and are committed to addressing it to improve the resilience and sustainability of their community. This collective approach fosters solidarity as they have come to recognise that climate issues cannot be solved in isolation. In this way, their awareness transcends immediate survival concerns and expands into a broader, more ethical understanding of coexistence with the natural world. This is the case of the Guna indigenous people being relocated to another land and questioning themselves about the future of their tradition and values, as well as of the Mongolian nomad shepherds realising how their landscape and own ancestral practices are bound to be impacted by a changing climate.

Even when individuals are not informed about the science of climate change, this sense of belonging to a wider community allows them to acknowledge the extent of humans' impact on the environment and the repercussions on communities beyond theirs. It is their experience that has made them very conscious of the relationship between human activity and the climate-related phenomena they have to deal with.

At the same time, recognising the different forms that climate change can take, the book goes beyond the narration of individuals' experiences to explain the varied climate change manifestations that populate their stories. In fact, once the spark of curiosity is ignited via these stories, we choose to pair them with relevant academic data and recent information in order to explain the reasons behind the phenomena described. This is crucial to better inform readers about the science underpinning these events, their connection with climate change, and their global significance, which extends far beyond the region described. Each of these stories, despite being profoundly local, is a first introduction to a global phenomenon, describing all the different ways in which our world is changing, and at a faster rate than we can imagine. From sea level rise, to landslides, floods, droughts and wildfires, the book portrays how climate is not only changing but also taking a direction which is pointing to the extreme. Extreme are the temperatures reached in certain areas of the world, and so are rainfall patterns and other weather events. The danger of reaching such extremes are the direct and indirect consequences faced daily by our planet, our ecosystems, natural resources, and ultimately people, as well as the ramifications they have on economy, health, culture, and traditions.

What emerges from the book, and in particular from the link between the stories and the description of global phenomena, is the need to address the existing disconnection between humans and the environment. For too long, the earth has been our playground as we have conceived ourselves and our history as separate from nature, rather than belonging within it. Such an erroneous human-nature divide, born out of values of dominance, clashes with our dependence on nature for our own survival. Often, this is what causes our incapacity to fully grasp the complexity of global challenges such as climate change.

This perception has determined the way we have built our civilisations, structured our economies, and even written our history – centring it on human achievements while often ignoring the larger ecological forces that have shaped, and continue to shape, our existence. However, human-made progress is but a fragment of a much broader and complex planetary story. The challenge lies in rebalancing this view to better integrate nature and our relationship with it.

Climate change is a stark reminder of this disconnect. While the topic is widely discussed, the way information is presented often alienates individuals rather than empowering them. Scientific data, though crucial, can be overwhelming and difficult to interpret. Catastrophic predictions can induce a sense of helplessness rather than action. Even as we witness climate-related events unfolding in real time – rising temperatures, extreme weather patterns, loss of biodiversity – many still find it difficult to connect these phenomena to their own lives in a tangible way. This dissociation, whether psychological or systemic, hinders our ability to respond effectively to what ultimately represents one of the most pressing challenges of our time.

For all these reasons, sometimes it might require us to exit our comfort zone and end up in remote corners of the world, such as the Erg Chigaga dunes, to finally understand and experience a sense of universal belonging. This was our experience and our starting point for looking for and telling these stories. Our hope is that this book will also serve as a journey in itself, taking you to remote and less-remote areas around the world, and in this way will introduce you to people who, just like many others, bridge this gap and heal modern society's divide to feel our way back into connection with nature.

By reframing our understanding in this way, we are called to confront the reality that our actions – individually and collectively – are deeply entwined with the health of the planet. Such a shift in perspective invites us to move beyond mere acknowledgment of the problem towards transformative action that respects the balance of ecosystems, fosters sustainable practices, and redefines our relationship with nature as one of stewardship, reciprocity, and mutual care.

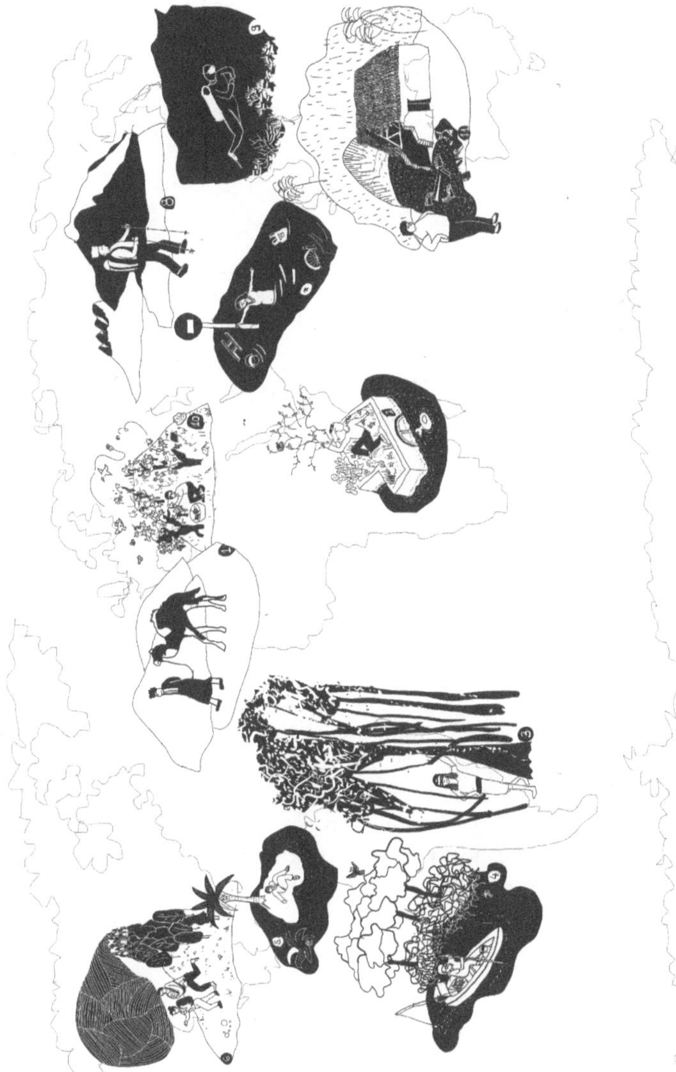

Figure 0.1 Map of featured stories across the world.

1 Settling down

"Look! There's a camel with blue eyes", said Brahim, walking through a sandy stretch, his wide blue scarf covering his face to show only his eyes and nose.

Brahim was born in the Sahara Desert, in the south of Morocco. He is an Arab nomad from a large family of five brothers and three sisters. Like every other nomad, he has two birthdays: the actual birth date and the moment when his family could register his birth in a nearby town.

The camel slowly makes its way towards us, a grin on its face.

> Camels with blue eyes are normally deaf, which means that mothers could lose their babies when adventuring into the desert. If this happens, they have a tacit agreement. Both the mother and the cub go to the last place where the mother had nursed the cub.

Brahim continues:

> But camels, our most precious resource, are becoming more and more rare. Camels, goats, and other animals are like currency for nomads. They are a source of transportation and provide milk, meat, and skins. If a family has a lot of camels, trains them well, and gets them to reproduce, they can always sell them on the market if they are in need of income. But now it's difficult and expensive to keep camels alive: families who used to have over one hundred barely have a dozen.

For centuries, the region where Brahim lives was a cross-cutting area of trade due to its main vital vein, the Drâa River. This river starts from the high Atlas Mountains and creates six oases, stretching over an area of over 200 kilometres. For centuries, Brahim's community demonstrated their capacity to adapt to tough environmental conditions also thanks to the relief offered by the river around this area. However, decreased rainfall and increasing droughts, coupled with the creation of a large dam in the area, have strongly affected the availability of water in the region.

DOI: 10.4324/9781003493662-2

"We did not see the problem with the dam at first nor with the changing climate", Brahim says.

> The real issues only happened after a few years. We started receiving water in the oases only once the dam was releasing it. The frequency was initially much higher, but now has diminished to only four times a year. And it really is not enough to keep plants and animals alive.

The construction of the El Mansour Eddahbi dam occurred in the 1970s in the proximity of Ouarzazate, aiming to generate hydroelectricity, facilitate irrigation, and regulate floods. Consequently, the downstream flow towards the nearby town of M'Hamid has diminished.

Nomads like Brahim and his family prepare for the summer months by digging large wells, sometimes in proximity to the river. "It used to be easier to find water in wells, but now I have to dig as deep as 16 or even 18 meters, and sometimes, the water is not there at all".

On top of increasing droughts and a dry river, Brahim explains how he himself has witnessed the changes in rainfall frequency. He says that 15 years ago, it used to rain between October and November, creating fertile lands and making plants flourish at the right time. However, now too little rain arrives too late, between January and February. This becomes an issue for animals and decimates even the most resilient ones. Reduced rainfall over the Atlas Mountains, which serve as the primary water sources for the river, has exacerbated conditions in the region.

The creation of the dam, together with more frequent droughts, are now forcing Brahim's family to adapt in a whole new way: settling down. Nomad's survival and culture were dependent on the presence of water across the region. Due to decreasing availability of water, animals slowly started to disappear. And without animals, nomads have little to sustain themselves.

"For thousands of years, we have adapted to the most difficult circumstances", Brahim says. "But now we need to adapt once again to something that goes against our very nature—moving to towns and living in the same place".

While Brahim is hopeful about his population's capacity to survive, he is also worried that precious stories from the past will disappear forever.

> Our tradition and lifestyle are now only sustained by tourism. But our ancient stories do not have any written form where they can be preserved for the future. The elders are now getting very old and start being unable to remember about our ancient ways of living.

For instance, he says, living in the desert taught him important wisdoms. He learnt to adapt to a simple life and become resilient to the hardest circumstances. This is a key feature of his community, and he is worried it will be lost with young people slowly moving to live in towns.

Living in the desert also made them creative, he argues, his eyes smiling in the sun. "When I was a kid, my mum used to invent beautiful stories to ease up the spirits in the desert's dire conditions". "You know, we get a lot of sandstorms here", he recalls. "And these can be very scary at first. You really can't see a thing and need to stand still". The first time he was caught in a sandstorm, Brahim's mum told him about a magic little prince who lived inside storms and who always brought good news. "'If you can keep your eyes open, look for the

Figure 1.1 Brahim El Gasmi, a Saharan nomad.

prince', she used to say, 'and ask him for a wish, anything you need'. Since then, I stopped being scared and started looking around as much as I could—and thirty years later, I'm still looking for this little prince".

Brahim's life is a colourful mix of these stories, beliefs, and traditions. Being a nomad for Brahim means adapting every day to a new scenario, a new challenge. And creativity is a key ingredient to his life. But things are changing: "My sisters have already moved and settled in the nearest town, M'Hamid", he points out with a note of resentment in his voice. As for him, he intends to keep a base in the town but still wanders around the desert as much as he can by driving curious tourists around.

A yellow dusty wind is blowing from the South. "It's getting tough, but we're powerful people", Brahim says. "We'll find a way, as we always have".

Across drylands: The global challenge of desertification

If you close your eyes and imagine a dryland, you are likely to picture a vast, apparently infinite, barren land where life is hardly imaginable. Yet, this is a common misconception, as drylands, which represent 41% of our world's surface land (Reid et al., 2005), are far more than just desolate deserts. Drylands include a variety of colourful landscapes, from arid and semi-arid lands to grasslands, savannahs, shrublands, and woodlands, and may therefore look quite different from that picture our mind is most likely directed to (IUCN, 2017).

Drylands are also far from being lifeless: they are home to more than 2 billion people as well as to a large variety of biodiversity (IUCN, 2017). But life in drylands is certainly different from other ecosystems. People, as well as animals, in these areas have had to learn how to adapt to extreme climatic conditions and variabilities, which are only getting worse and more uncertain due to climate and land use change. For instance, increasing evaporation may lead to less rain which subsequently impacts the water supply for both vegetation and soils. Additionally, reduced rainfall can lead to completely dried-out soils, making them more susceptible to erosion (Mirzabaev et al., 2019). Further, deforestation, resulting from human activities such as overgrazing and unsustainable agricultural practices, is one of the key enemies of drylands. Without vegetation to anchor the soil, drylands become more susceptible to winds and water, exacerbating desertification. Extractive industries in drylands can also alter soil composition and the availability of water. And water availability in these ecosystems is already well below the recommended threshold[1] (FAO, 2016).

Among the different types of drylands, deserts are certainly the most remarkable and where life hits hardest. But deserts are not just the hot, sandy stretches of land we have all heard of, such as the Sahara in North Africa. In fact, deserts can also be cold, like the Asian Gobi Desert, mountainous, or even ice-covered such as the Antarctic Polar Desert in Antarctica. The common denominator of deserts, as drylands, remains the low frequency and scarcity of rainfall (less than 25 centimetres per year) (Crawford & Gosz, 1982).

Deserts cover one-fifth of the Earth's land surface, but these areas are increasing at an alarming rate.

When drylands degrade, they progressively desertify. While there is no universally agreed upon definition of desertification, this is intended as land degradation in arid, semi-arid, and dry sub-humid areas resulting from many factors, including climatic variations and human activities (UNCCD, 1994). This phenomenon has become a common and increasing threat, with approximately one-third of drylands' surface being already degraded (McSweeney, 2019). One of the major causes and outcomes of such degradation is biodiversity loss. Beyond vegetation and endemic species, dryland biodiversity includes soil organisms that have uniquely adapted to the conditions of this ecosystem. Losing such soil biodiversity will inevitably impact the sustainability of the soil itself (IUCN, 2017).

Desertification is already directly affecting 250 million people, but putting at risk many more, due to implications on food security, livelihoods, and health (McSweeney, 2019). However, expectedly, these challenges are affecting the population disproportionately, with 85% of impacted people living in poorer – developing or newly industrialised – nations (Burrell et al., 2020).

Desertification is increasing the surface of existing deserts, with many semi-arid regions turning into real deserts (Nunez, n.d.). It is also contributing to climate change, as dryland degradation is releasing a substantial part of stored carbon back to the atmosphere, along with other greenhouse gases, such as nitrous oxide (Davies, 2017). According to climate scientists, drylands will expand by up to 10% by the end of the 21st century[2] with around 80% of them growing in developing countries (IUCN, 2017). Loss of vegetation and ecosystem disruptions due to changes in weather patterns are likely to make such drylands progressively inhabitable.

Existing deserts are equally becoming less inhabitable due to climate and land use change. Storms of sand and dust have increased, leading to an estimated 25% increase in dust emissions compared to the late 19th century (Stanelle et al., 2014). These so-called "white storms", as suggestive as they might seem, are affecting local air quality and posing a potential health risk. Further, deserts are adversely affected by less predictable rainfall patterns and overheating as their fragile ecosystems are not ready to manage sudden changes. This includes its species who are highly specialised to be able to survive in such a harsh environment; however, they are thus particularly sensitive to any fluctuation in temperature and rainfall.

Nowadays, human communities living in drylands find themselves at a crossroad. Traditional livelihoods linked to farming and herding are disrupted, and migration, often to urban areas, becomes a survival strategy. While dryland populations have significant past experience and sources of resilience to successfully adapt to climatic changes, they are also highly vulnerable to desertification and climate change (Mirzabaev et al., 2019). Pastoral and agropastoral households in drylands are among the most vulnerable. Estimates suggest that 100 to 200 million people practice pastoralism globally, with 30 to 63 million being nomadic (Dong et al., 2016).

In particular, changing climates have a harsh effect on nomadic communities pursuing livelihoods who are heavily dependent on nature. These communities, while traditionally presenting a strong adaptive capacity to dryland climate variability, are also increasingly facing land degradation, which impacts the viability of herding, leading in turn to conflicts and marginalisation (Varghese & Singh, 2016). Consequently, pastoral communities are ill-prepared for climate extremes and are among the most food-insecure groups globally (Dong et al., 2016; Quevedo et al., 2019). In response to these strategies, nomadic herders have developed a range of adaptation strategies involving solutions such as moving to areas with better water and pasture, improving seasonal access to water and feed, shifting herds' composition and, at last, diversifying livelihoods (Tugjamba et al., 2023).

The latter includes exploring alternative income sources, such as handcrafts, to reduce reliance on livestock alone. At the same time, such strategies require access to markets, and often education, which is not easy to ensure in certain drylands. Moreover, integration of new practices must align with traditional knowledge and cultural values to ensure community acceptance and sustainability. Such knowledge encompasses not only an understanding of animal husbandry but also insights into plant species, weather patterns, and the ecological relationships within their environments. The preservation of this knowledge is crucial as it forms the foundation upon which communities can build resilience against the impacts of climate change. Efforts to document and support traditional practices can empower nomadic communities, allowing them to adapt to new realities while maintaining their cultural heritage.

Notes

1 This refers to the minimum threshold to ensure human well-being and sustainable development.
2 This number is based on a high greenhouse gas (GHG) emissions scenario.

Bibliography

Burrell, A. L., Evans, J. P., & De Kauwe, M. G. (2020). Anthropogenic Climate Change Has Driven Over 5 Million km2 of Drylands Towards Desertification. *Nature Communications*, 11(1), 3853. https://doi.org/10.1038/s41467-020-17710-7

Crawford, C. S., & Gosz, J. R. (1982). Desert Ecosystems: Their Resources in Space and Time. *Environmental Conservation*, 9(3), 181–195.

Davies, J. (2017). The Land in Drylands: Thriving in Uncertainty Through Diversity. *Global Land Outlook, Working Paper*, United Nations Convention to Combat Desertification (UNCCD). Available from: https://www.unccd.int/sites/default/files/2018-06/15.%20The%2BLand%2Bin%2BDrylands__J_Davies.pdf [Last accessed: 29th November 2024].

Dong, S., Kassam, K. A. S., Tourrand, J. F., & Boone, R. B. (2016). *Building Resilience of Human-Natural Systems of Pastoralism in the Developing World.* Springer.

FAO (2016). *Trees, Forests and Land Use in Drylands: The First Global Assessment.* Food and Agriculture Organization of the United Nations (FAO). Available from: https://openknowledge.fao.org/server/api/core/bitstreams/587c8b7d-caa8-45d2-bcba-ec2ffb5c2143/content [Last accessed: 29th November 2024].

IUCN (2017). *Drylands and Land Degradation.* International Union for Conservation of Nature. Available from: https://iucn.org/sites/default/files/2022-07/drylands_and_land_degradation_issues_brief_0.pdf [Last accessed: 29th November 2024].

McSweeney, R. (2019). Explainer: 'Desertification' and the Role of Climate Change. *Carbon Brief.* Available from: https://www.carbonbrief.org/explainer-desertification-and-the-role-of-climate-change/

Mirzabaev, A., Wu, J., Evans, J., García-Oliva, F., Hussein, I. A. G., Iqbal, M. H., Kimutai, J., Knowles, T., Meza, F., Nedjraoui, D., Tena, F., Türkeş, M., Vázquez, R. J., & Weltz, M. (2019). Desertification. In: P. R. Shukla, J. Skea, E. Calvo Buendia, V. Masson-Delmotte, H.-O. Pörtner, D. C. Roberts, P. Zhai, R. Slade, S. Connors, R. van Diemen, M. Ferrat, E. L. Haughey, S., S. Neogi, M. Pathak, J. Petzold, J. Portugal Pereira, P. Vyas, E. Huntley, K. Kissick, M. Belkacemi, & J. Malley (Eds.) *Climate Change and Land: An IPCC Special Report on Climate Change, Desertification, Land Degradation, Sustainable Land Management, Food Security, and Greenhouse Gas Fluxes in Terrestrial Ecosystems.* IPCC. https://doi.org/10.1017/9781009157988.005

Nunez, C. (n.d). Deserts, Explained. *National Geographic.* Available from: https://www.nationalgeographic.com/environment/article/deserts [Last accessed: 29th November].

Quevedo, R., Milán, M., & López-i-Gelats, F. (2019). Pathways of Incorporation of Young Farmers into Livestock Farming. *Land Use Policy.* https://doi.org/10.1016/j.landusepol.2019.03.052

Reid, W., Mooney, H., Cropper, A., Capistrano, D., Carpenter, S., & Chopra, K. (2005). *Millennium Ecosystem Assessment. Ecosystems and Human Well-Being: Synthesis.* World Resources Institute.

Stanelle, T., Bey, I., Raddatz, T., Reick, C., & Tegen, I. (2014). Anthropogenically Induced Changes in Twentieth Century Mineral Dust Burden and the Associated Impact on Radiative Forcing. *Journal of Geophysical Research: Atmospheres*, 119(23), 13, 526–513, 546. https://doi.org/10.1002/2014JD022062

Tugjamba, N., Walkerden, G., & Miller, F. (2023). Adapting Nomadic Pastoralism to Climate Change. *Climatic Change*, 176(4), 28. https://doi.org/10.1007/s10584-023-03509-0

UNCCD (1994). *General Assembly. Elaboration of an International Convention to Combat Desertification in Countries Experiencing Serious Drought and/or Desertification, Particularly in Africa.* United Nations Convention to Combat Desertification. Available from: https://www.unccd.int/sites/default/files/relevant-links/2017-01/English_0.pdf [Last accessed: 9th May 2025].

Varghese, N., & Singh, N. P. (2016). Linkages between Land Use Changes, Desertification and Human Development in the Thar Desert Region of India. *Land Use Policy*, 51, 18–25. https://doi.org/10.1016/j.landusepol.2015.11.001

2 A village underwater

"Some stories are happening under our eyes, just as we tell them. But once we realise that our words have become reality, it's often too late to do something about it". Robert talks to us with an alarmed voice, walking through a set of wooden structures which he and his village peers have improvised as emergency homes while their entire village is completely underwater, only the roofs visible from the lake.

Jonathan Karigu Gachau, also known as Robert for his friends, lives in the village of Kamere, on the side of Lake Naivasha, one of the many lakes nestled within the Great Rift Valley in Kenya. Unlike other Rift Valley lakes, which are typically saline, Lake Naivasha is a freshwater oasis. The area around the lake, fringed with a dense growth of papyrus and acacia trees, is a fertile ground for agriculture, becoming a hub for Kenya's floriculture industry exported worldwide, and attracting a wide number of wild animals who go there to satisfy their thirst.

Robert's main occupation traditionally has been to take tourists for "lake safaris", a very common activity in the area. He has been working in this field since 2012. Tourists frequently stop by this lake on their way to the nearby Masai Mara Reserve to jump on a long pagoda and observe zebras, giraffes, and buffalos drinking water at the shore, and waterbucks and hippos swimming all around the lake.

However, Lake Naivasha today is not the same safari paradise that used to be. From a satellite, one might not realise anything as the lake has just grown bigger. However, once one arrives in the proximity of the lake, strange branches spurring from the water and floating roofs show how it has progressively submerged entire villages in its surroundings.

"The water from the lake started rising a long time ago, already in 2016", Robert tells us. "During Covid times, in 2020, the water level had already risen quite a few meters. The first thing that shocked me about this was to see so many acacia trees dying around Naivasha, as they could not breathe any longer".

At the time, villagers were concerned about the lake water having started to reach their houses and damaging some of the supplies. However, everyone was convinced that the situation would soon return to normal, and no major

DOI: 10.4324/9781003493662-3

precautions were taken. "There was a lot of rain during the rainy season, more than we could ever witness. But no one would have ever thought that the rain would be enough for the lake to overflow, completely covering our village".

The second phenomenon that shocked Robert was to witness the hippos in the lake trying to eat the water hyacinths. "As the grass was being submerged, hippos were becoming hungrier and ended up eating something different: the hyacinths forming in the water, including their flower. It was the first time I saw them so desperate, so aggressive", he tells us.

During the same period, a group of environmentalists and conservationists had arrived in the area and tried to study the phenomenon. "I was asked to take these people along the Malewa river, the main tributary that flows into Lake Naivasha. What was clear during that trip was that the quality of the water of the Malewa river was slowly deteriorating, becoming more brown and less transparent than usual. We also saw a large number of people cutting down trees for charcoal production".

The scientists told Robert that heavy rains, together with deforestation in the area, was likely causing silting, and the new sediments settled at the bottom were in turn lowering the depth of the lake.

However, this was not an isolated phenomenon: most of the other lakes in the Rift valley had started to overflow too, albeit each one to a different extent. "We started hearing similar stories from our friends, our neighbours, and other families living around lakes Turkana, Nakuru, Bogoria and Baringo", he continues. "Baringo, in particular, was where the phenomena was the worst. There, the hyacinth had started infesting the water that had overflown into their villages. While everyone was talking about Baringo, we were not realising that the water that surrounded us had gotten progressively green - the same was happening to us".

About eight hundred families in Robert's village now have their homes under water, covered by a thick layer of green hyacinth. This means that only the roofs can be seen from the shore. The water, in turn, is attracting a lot of mosquitos, thereby worsening the health conditions in the village.

> Some people tried to do something to prevent the silent but steady rising of the lake water. A Dutch flower farmer living nearby Kamere decided to build a dam in front of his fields, which was 3 meters high. But he underestimated how high the water was going to rise, and he also did not fully grasp the strength of the water, which ended up breaking the barriers.

Robert also tells us how many people working in tourism, such as the ones owning small hotels and campsites around the lake, lost their jobs as these structures were completely submerged. Most of these people fled, and the few that remained needed to find another way to make ends meet around the lake. Of these, many resorted to fishing. "However, you need to have a lot of experience to be a fisherman here", Robert tells us. Fishing is a dangerous activity on this lake as it requires a very good understanding of the hippos to prevent them from attacking humans.

Robert himself is now struggling as tourists are much less numerous. "Obviously I still manage to do boat tours, as the water is not what's missing," he says.

> People are still coming around, and sometimes they do not even realise what happened unless I tell them. Sometimes I am surprised, but then I realise that if you don't know what is buried underwater you might not understand how the lake used to look like before.

His family has also been heavily impacted financially.

> Last month, we needed to rent a new house as we were left with no other choice but to abandon our own. My wife, who was growing carrots and other vegetables in our own garden and selling them at the local market, has now lost her job, and is now looking for another one in a nearby flower farm.

At the same time, nobody knows when the water will stop overflowing. Robert tells us that the water is now also getting very close to the new place they're living in.

> I do not know what I will do if I need to move out of this new rental place. Every day we continue to experience a lot of rainfall, even if it's not the rainy season anymore. This makes me very worried, as we cannot truly forecast when the rain will stop anymore.

We ask Robert if he and his village peers still believe that the situation will go back to normal. "This is what the government is telling us, as they just ordered us to vacate our homes without providing any long-term solutions", he says.

> But a large part of the village still hopes to be able to go back to their homes, perhaps when the weather gets drier. On my end, while I am trying to stay positive, I understand that the world is changing, and it will be almost impossible to go back.

Robert believes that new solutions will need to be found.

> I am not an architect, but I am convinced that we should explore options such as houses built on high poles, to cater for possible lake level rise. And we need more research into this phenomenon, to be able to assess how much it will worsen in the future. We all need to be prepared, as the world is changing in front of our eyes, and is doing so as we speak.

Figure 2.1 Jonathan Karigu Gachau, tourist guide, standing in front of Lake Naivasha.

Shaping lakes: An undervalued ecosystem under increasing threats

If one were to look at a map of the earth, bodies of water known as lakes and ponds of all shapes and sizes would appear, scattered across different regions, and types of landscapes, from mountains to valleys and cities. Yet, at first glance, one might not realise that the geography of lakes has changed and continues to do so at a slow, but persistent pace.

Historically, lakes have been at the backbone of human civilisation, providing both food and water supply, as well as the means for travel and transportation. Since the beginning of human development, lakes have gone a long way and established themselves as essential parts of the world's geography and functions, even though their ecological significance has been traditionally undervalued. This is due to the challenges in valuing ecosystem services as well as the lack of sufficient data on lake biodiversity which contributes to the complexity of this ecosystem (Xie et al., 2024).

While accounting for only 3% of the Earth's surface (Downing et al., 2006) and for 21% of terrestrial aquatic ecosystems (Xie et al., 2024), lakes represent a crucial part of the water cycle, being able to collect water in a limited area (Rutledge et al., 2023). In fact, the water resources stored by lakes are almost 90 times higher than those stored by rivers (Xie et al., 2024). People, animals, and plants all largely depend on these freshwater ecosystems. The over 100 million lakes that populate our planet (Verpoorter et al., 2014) store almost 90% of the earth's liquid surface freshwater, providing a crucial source of drinking water to surrounding communities (Woolway et al., 2022; Yao et al., 2023). Water is also used for irrigation and fisheries as well as power generation, stressing lakes' contribution to food and energy security (Woolway et al., 2022). This important role, combined with their attractive landscape, makes lakes an important touristic attraction and a traditionally appreciated space for recreation, therefore also contributing to employment opportunities (Zhang et al., 2024). The dependency of people on these bodies of water is often undervalued, but it is far stronger than what we tend to believe. For instance, Lake Victoria, the biggest in the African continent, counts 42 million people relying on its services, as their primary source of drinking water, food, and employment (Nyamweya et al., 2023). Similarly, the Great Lakes in North America have been estimated to lead to the creation of more than 1.3 million jobs (Rau et al., 2020).

Plants and animals have also made lakes their home. In fact, the diversity of species populating all freshwater ecosystems – including lakes, rivers, streams, and wetlands – is considerably higher than in other ecosystems (IUCN, 2008). However, freshwater biodiversity is also declining faster (Lynch et al., 2023; WWF, 2020) as human pressures and climate change are increasingly putting these ecosystems, and their inhabitants, at risk (IUCN, 2008). Some threats such as eutrophication – whereby increasing nutrients in the water lead to denser plant growth – are also expected to worsen with climate change, further contributing to freshwater biodiversity decline (Rodgers, 2021).

Further, lakes are also important contributors to global carbon cycles. More than half of the carbon that lakes receive from the watershed is buried or returned to the atmosphere. As such, lakes are important regulators of climate (Williamson et al., 2009; Zwart et al., 2019). For instance, they act as heat sinks in the warmer periods and heat sources during colder ones, providing natural infrastructures capable of mitigating droughts and controlling floods (Zhang et al., 2024). As lakes are particularly responsive to variations in climate, they have become important sentinels of climate change (Woolway et al., 2022). However, such a strong link between these ecosystems and climate also makes them particularly vulnerable to both natural stressors and human-induced pressures. In fact, these factors have contributed to the deterioration of lakes around the world, leading to the loss of biodiversity and key ecosystem services (Pi et al., 2022; Xie et al., 2024).

Climate variations directly or indirectly affecting lakes include extreme events, alongside changes in temperature and precipitation. For instance, higher temperatures are found to increase water temperatures as well. The warming rate of lakes' water around the world has been observed at 0.34 degrees Celsius per decade, with scientists pointing to a greater warming rate than that of either the ocean or the atmosphere (O'Reilly et al., 2015). The year 2022 was the warmest year on record for lakes globally (Copernicus Climate Change Service, 2021). Yet, lake water temperature is expected to increase even more, by 1–4 degrees Celsius by the end of the century (Woolway et al., 2022). The result of increasing water temperatures is lower ice cover and higher evaporation, which is expected to increase by 16% annually by the end of the century (Wang et al., 2018), resulting in lower lake water levels (Woolway et al., 2022). This phenomenon has indeed materialised in many natural lakes where water levels continue to decline progressively. The impact of a changing climate together with increased water consumption are likely to make this trend continue over time (Ozdemir et al., 2023).

On the other hand, frequent and intense precipitations as well as melting glaciers seem to contribute to the opposite effect: rising lake water levels (Kuraji & Saito, 2022). Several lakes around the world have experienced this phenomenon, to the point of their shape changing drastically, submerging nearby villages, and ultimately changing geographies (Pi et al., 2022).

In 2019, the second largest lake known on Earth – Lake Victoria, shared between Kenya, Uganda, and Tanzania – started to rise, reaching record-breaking levels in mid-May 2020. In the same years, the area around the lake in Kenya had witnessed intense rainfall, which led to floods and the consequent displacement of thousands of people. When the Nzoia River, in the Lake Victoria basin, overflew, 40,000 people were displaced (Pietroiusti et al., 2024). Episodes of this kind, while anomalous in nature, are actually not new. On the contrary, already throughout the 2010s, rising water levels trends were observed in several lakes located in Kenya: Lake Bogoria, Naivasha, Nukuru, as well as Turkana – the largest desert lake on earth – all flooded. The magnitude of the issue continued to rise slowly, reaching worrying levels in recent years with nearly 400,000 Kenyans displaced. In fact, since 2010, the yearly average rainfall in the country was higher than ever and in 2019, Kenya was subject to the third highest amount of rainfall it had ever recorded (Baraka, 2022).

The same trend has been experienced by lakes in other regions of the world. For instance, the largest freshwater lake in China, Poyang, has recently been witnessing water level rise following heavy precipitation. While the lake naturally fluctuates in size between winter and summer, it reached unusually high water levels between 2019 and 2020 (Guo et al., 2022; Hansen, 2020), and more recently, in 2024, the area of the lake expanded by 1,000 square kilometres within a single month (Xinhua, 2024). Similarly, an analysis of the state and trends of Lake Champlain, located in North America, showed how intense rainstorms in 2023 led to increased lake levels, a phenomenon already experienced in 2011 when severe floods took place in the basin area (Lake Champlain Basin Program, 2024).

While alternation in rainfall is recognised as a result of climate change, it is important not to underestimate the role played by other stressors as well as human pressures, resulting in lake overflows. For instance, landslides and water runoff may result from storms and precipitation as well as from deforestation and other land use changes, similarly contributing to rising lake water levels (Baraka, 2022). In fact, deforestation and agricultural practices, by affecting soil directly, contribute to its destabilisation, making erosion more likely and reducing infiltration, thereby increasing flood risks when water levels are high (Bradshaw et al., 2007; Kimtai et al., 2024).

What is shaping our lakes may be hard to grasp, as the observed changing lake dynamics around the world point to both an underestimation and limited understanding of hydrological impacts and of the vulnerability of lakes to different stressors (Kuraji & Saito, 2022; Torabi Haghighi & Kløve, 2015). At the same time, with lake water levels simultaneously rising and falling in different parts of the globe, it becomes clear that these important, yet undervalued ecosystems are indeed changing drastically in many cases at the expenses of people, plants, and animals. Ensuring that lakes continue to fulfil their ecological, economic, and social functions calls for stronger efforts to fill the apparent knowledge gap, while also exploring innovative solutions to reduce the cascading effects of changing lake water levels, preserving biodiversity in the areas and ensuring accurate land use management, ultimately increasing the resilience of these ecosystems to changes in climate.

Bibliography

Baraka, C. (2022). A Drowning World: Kenya's Quiet Slide Underwater. *The Guardian*. https://pulitzercenter.org/stories/drowning-world-kenyas-quiet-slide-underwater [Last accessed: 5th December 2024].

Bradshaw, C. J. A., Sodhi, N. S., Peh, K. S.-H., & Brook, B. W. (2007). Global Evidence That Deforestation Amplifies Flood Risk and Severity in the Developing World. *Global Change Biology*, 13(11), 2379–2395. https://doi.org/10.1111/j.1365-2486.2007.01446.x

Copernicus Climate Change Service. (2021). *European State of the Climate 2021: Full report. European Centre for Medium-Range Weather Forecasts*. Available from: https://climate.copernicus.eu/esotc/2021 [Last accessed: 9th May 2025].

Downing, J., Prairie, Y., Cole, J., Duarte, C., Tranvik, L., Striegl, R., McDowell, W., Kortelainen, P., Caraco, N. F., Melack, J. M., & Middelburg, J. (2006). The Global Abundance and Size Distribution of Lakes, Ponds, and Impoundments. *Limnology and Oceanography*, 51, 2388–2397. https://doi.org/10.4319/lo.2006.51.5.2388

Guo, Y., Lai, X., Wu, X., & Chen, Y. (2022). Dry-Season Water Level Shift Induced by Channel Change of the River-Lake System in China's Largest Freshwater Lake, Poyang Lake. https://doi.org/10.21203/rs.3.rs-1751827/v1

Hansen, K. (2020). Poyang Lake Extremes. *NASA*. Available from: https://earthobservatory.nasa.gov/images/146987/poyang-lake-extremes [Last accessed: 29th November].

IUCN (2008). *Freshwater Biodiversity - A Hidden Resource Under Threat*. International Union for Conservation of Nature. Available from: https://iucn.org/sites/default/files/import/downloads/freshwater_biodiversity_a_hidden_resource_under_threat_factsheet_en.pdf [Last accessed: 5th December 2024].

Kimtai, D. J., Makokha, G. O., & Sichangi, A. W. (2024). Modeling of Water Level Trends and Characterizing Potential Influencing Factors in Lake Baringo in Kenya. *Discover Water*, 4(1), 55. https://doi.org/10.1007/s43832-024-00105-w

Kuraji, K., & Saito, H. (2022). Long-Term Changes in Relationship between Water Level and Precipitation in Lake Yamanaka. *Water*, 14(14), 2232.

Lake Champlain Basin Program (2024). *2024 Lake Champlain State of the Lake and Ecosystem Indicators Report*. Lake Champlai Basin Program. Available from: https://www.lcbp.org/wp-content/uploads/2024/06/2024-State-of-the-Lake-Report.pdf [Last accessed: 5th December 2024].

Lynch, A. J., Cooke, S. J., Arthington, A. H., Baigun, C., Bossenbroek, L., Dickens, C., Harrison, I., Kimirei, I., Langhans, S. D., Murchie, K. J., Olden, J. D., Ormerod, S. J., Owuor, M., Raghavan, R., Samways, M. J., Schinegger, R., Sharma, S., Tachamo-Shah, R.-D., Tickner, D., & Jähnig, S. C. (2023). People Need Freshwater Biodiversity. *WIREs Water*, 10(3), e1633. https://doi.org/10.1002/wat2.1633

Nyamweya, C., Lawrence, T. J., Ajode, M. Z., Smith, S., Achieng, A. O., Barasa, J. E., Masese, F. O., Taabu-Munyaho, A., Mahongo, S., Kayanda, R., Rukunya, E., Kisaka, L., Manyala, J., Medard, M., Otoung, S., Mrosso, H., Sekadende, B., Walakira, J., Mbabazi, S., & Nkalubo, W. (2023). Lake Victoria: Overview of Research Needs and the Way Forward. *Journal of Great Lakes Research*, 49(6), 102211. https://doi.org/10.1016/j.jglr.2023.06.009

O'Reilly, C. M., Sharma, S., Gray, D. K., Hampton, S. E., Read, J. S., Rowley, R. J., Schneider, P., Lenters, J. D., McIntyre, P. B., Kraemer, B. M., Weyhenmeyer, G. A., Straile, D., Dong, B., Adrian, R., Allan, M. G., Anneville, O., Arvola, L., Austin, J., Bailey, J. L., & Zhang, G. (2015). Rapid and Highly variable Warming of Lake Surface Waters Around the Globe. *Geophysical Research Letters*, 42(24), 10), 773–710, 781. https://doi.org/10.1002/2015GL066235

Ozdemir, S., Yaqub, M., & Yildirim, S. O. (2023). A Systematic Literature Review on Lake Water Level Prediction Models. *Environmental Modelling & Software*, 163, 105684. https://doi.org/10.1016/j.envsoft.2023.105684

Pi, X., Luo, Q., Feng, L., Xu, Y., Tang, J., Liang, X., Ma, E., Cheng, R., Fensholt, R., Brandt, M., Cai, X., Gibson, L., Liu, J., Zheng, C., Li, W., & Bryan, B. A. (2022).

Mapping Global Lake Dynamics Reveals the Emerging Roles of Small Lakes. *Nature Communications*, 13(1), 5777. https://doi.org/10.1038/s41467-022-33239-3

Pietroiusti, R., Vanderkelen, I., Otto, F. E. L., Barnes, C., Temple, L., Akurut, M., Bally, P., van Lipzig, N. P. M., & Thiery, W. (2024). Possible Role of Anthropogenic Climate Change in the Record-Breaking 2020 Lake Victoria Levels and Floods. *Earth System Dynamics*, 15(2), 225–264. https://doi.org/10.5194/esd-15-225-2024

Rau, E., Riseng, C., Vaccaro, L., & Read, J. G. (2020). *The Dynamic Great Lakes Economy - Employment Trends from 2009 to 2018*. Michigan Sea Grant. Available from: https://www.michiganseagrant.org/wp-content/uploads/2021/03/MICHU-20-203-Great-Lakes-Jobs-Report.pdf [Last accessed: 10th December 2024].

Rodgers, E. M. (2021). Adding Climate Change to the Mix: Responses of Aquatic Ectotherms to the Combined Effects of Eutrophication and Warming. *Biol Lett*, 17(10), 20210442. https://doi.org/10.1098/rsbl.2021.0442

Rutledge, K., McDaniel, M., Teng, S., Hall, H., Ramroop, T., Sprout, E., Hunt, J., Boudreau, D., & Costa, H. (2023). Lake - Dictionary Entry. In J. Evers, E. Editing, & K. West (Eds.). National Geographic Society. Available from: https://education.nationalgeographic.org/resource/lake/ [Last Accessed: 9th May 202].

Torabi Haghighi, A., & Kløve, B. (2015). A Sensitivity Analysis of Lake Water Level Response to Changes in Climate and River Regimes. *Limnologica*, 51, 118–130. https://doi.org/10.1016/j.limno.2015.02.001

Verpoorter, C., Kutser, T., Seekell, D. A., & Tranvik, L. J. (2014). A Global Inventory of Lakes Based on High-Resolution Satellite Imagery. *Geophysical Research Letters*, 41(18), 6396–6402. https://doi.org/10.1002/2014GL060641

Wang, W., Lee, X., Xiao, W., Liu, S., Schultz, N., Wang, Y., Zhang, M., & Zhao, L. (2018). Global lake Evaporation Accelerated by Changes in Surface Energy Allocation in a Warmer Climate. *Nature Geoscience*, 11(6), 410–414. https://doi.org/10.1038/s41561-018-0114-8

Williamson, C., Saros, J., Vincent, W., & Smol, J. (2009). Lakes and Reservoirs as Sentinels, Integrators, and Regulators of Climate Change. *Limnology and Oceanography*, 54. https://doi.org/10.4319/lo.2009.54.6_part_2.2273

Woolway, R. I., Sharma, S., & Smol, J. P. (2022). Lakes in Hot Water: The Impacts of a Changing Climate on Aquatic Ecosystems. *Bioscience*, 72(11), 1050–1061. https://doi.org/10.1093/biosci/biac052

WWF (2020). *Living Planet Report 2020 - Bending the Curve of Biodiversity Loss*. World Wildlife Fund. Available from: https://www.wetlands.org/wp-content/uploads/2020/09/lpr-2020-full-report.pdf [Last accessed: 9th May 2025]

Xie, H., Ma, Y., Jin, X., Jia, S., Zhao, X., Zhao, X., Cai, Y., Xu, J., Wu, F., & Giesy, J. P. (2024). Land Use and River-Lake Connectivity: Biodiversity Determinants of Lake Ecosystems. *Environmental Science and Ecotechnology*, 21, 100434. https://doi.org/10.1016/j.ese.2024.100434

Xinhua. (2024). China's Largest Freshwater Lake Sees Rising Water Level. *China Daily*. Available from: https://www.chinadaily.com.cn/a/202404/07/WS66124bfda31082fc043c08c6.html [Last accessed: 10th December 2024].

Yao, F., Livneh, B., Rajagopalan, B., Wang, J., Crétaux, J.-F., Wada, Y., & Berge-Nguyen, M. (2023). Satellites Reveal Widespread Decline in Global Lake Water Storage. *Science*, 380(6646), 743–749. https://doi.org/doi:10.1126/science.abo2812

Zhang, Y., Deng, J., Zhou, Y., Zhang, Y., Qin, B., Song, C., Shi, K., Zhu, G., Hou, X., Zhang, Y., He, S., Woolway, R. I., & Li, N. (2024). Drinking Water Safety Improvement and Future Challenge of Lakes and Reservoirs. *Science Bulletin*, 69(22), 3558–3570. https://doi.org/10.1016/j.scib.2024.06.018

Zwart, J. A., Hararuk, O., Prairie, Y. T., Jones, S. E., & Solomon, C. T. (2019). Improving Estimates and Forecasts of Lake Carbon Dynamics Using Data Assimilation. *Limnology and Oceanography: Methods*, 17(2), 97–111. https://doi.org/10.1002/lom3.10302

3 Green lungs in peril

Vanda Ortega Witoto is sitting on a wooden bench outside of her small house on the banks of the Solimões River, an upper stretch of the Amazon river which conflates with the Rio Negro, in the very depth of the Brazilian Amazons. Behind her, a bright palm tree is swaying gently in the warm, humid breeze of the afternoon.

"Finding who I really am has taught me what the forest truly means for me, and for my community" she says, looking at the tall palm trees. Vanda belongs to the Witoto tribe, a South American indigenous community originating from southeastern Colombia and northern Peru. In Brazil, her community is living within the Tribe Park in Manaus, the first recognised indigenous neighbourhood.

However, Vanda lived for a long time far away from her community. "At the age of 16, I was forced to leave my family to move to the nearest city, Manaus, and work as a housekeeper. This is due to the poverty that many indigenous people still face".

The city was a completely different setting and Vanda found herself pushed into a life of hard work, where she slowly became used to following orders and rarely had time to think about her past. "At the same time, I was curious about the world around me, so I decided to study in parallel and finish high school" she tells us, her eyes filled with pride.

One day, when talking to a friend, she learned that universities were giving free courses to students with indigenous backgrounds. This opened up an entire new world for her, and she decided that she wanted to give it a try. "Obviously, the first step was to obtain a certificate that could prove my indigenous heritage" she says. "This is when I started asking myself questions about the Witoto, as I never understood who we fully were". Delving into her roots, she began to understand the significance of her heritage. She learned about colonialism and its devastating impact on indigenous people, and this newfound knowledge ignited a passion for activism within her.

DOI: 10.4324/9781003493662-4

"As I started my studies, I became fuelled with the desire to preserve the language and tradition of my community, against the strong racism faced by my people and the systematic efforts to erase our cultural identity".

> I learned about phenomena such as extreme droughts, forest fires, extreme heat impacting our agriculture – all of them were caused by the standards of living of the rest of the world, and yet, my people were amongst the ones suffering the most.
>
> Indigenous communities, with their customs and traditions, have always lived in harmony with nature. For us, it also has a spiritual meaning, in addition to representing a crucial aspect of our subsistence. But the Amazon forest that surrounds us has been suffering more and more from the changes in climate and human activities, and so have we.

The Amazon forest plays a crucial role in absorbing carbon dioxide from the atmosphere, but its progressive degradation is exacerbating rather than reducing global warming. Witoto people, lacking basic rights such as education and access to clean water, have for a long time been marginalised, while their territory was progressively conquered by land grabbers, interested in illegal mining and deforestation, and the consequences of climate change were more and more felt by her people.

"My people have fought long to obtain demarcated territories, where they could preserve their livelihood and culture" she explains. However, while physical demarcation was fundamental to recognise their rights and to keep off illegal miners, it was not enough to protect them from climate and environmental degradation.

One concrete example is when, in 2023, the region experienced the worst drought in history: at the time. The largest rivers in the Amazon Basin reached record drought levels, leaving 62 municipalities in a state of emergency.

> During this time, we also had the third worst air quality in the world, due to forest fires, some illegal and man-made, and others driven by the El Niño climate phenomenon. The smoke was suffocating, you could feel it in your skin. It trapped the heat, our eyes were burning, and even drinking water didn't help.

The health of all of Vanda's community, the Witoto people, was impacted during this time.

"Another example is to be found in the damages that climate change has caused to agriculture in the area. Witoto people used to live in full harmony with nature and be completely self-sufficient". However, nowadays this is not possible anymore. "Harvesting is not the same as it used to be. We see it in every fruit that we harvest", Vanda tells us.

> For instance, the bananas that grow today are not the same in quantity and quality, as the soil is too warm. People also used to know what season

> to plant and harvest, but now it is much less clear due to extreme events and seasonal changes.

All of these cause losses, hunger, and poverty due to the unpredictability of the climate.

Vanda's voice breaks when she talks about the desperation reached by her people due to malnourishment. She continues

> I see people in such extreme poverty that they are willing to trade women or children to get basic access to food. And they often receive from the government a plastic basket with food that is not even from this region, with foreign ingredients made elsewhere in the world. All of this is so wrong.

Food insecurity is paired with issues of access to water. "These stories become even more of a paradox when living next to the Amazon river, the longest freshwater body in the world. Yet, one of the largest public health problems we are facing is access to uncontaminated water." In fact, elevated mercury in indigenous people is causing multiple health problems across Vanda's tribe, including impairing women's ability to get pregnant and causing deformities in children.

When asked about how things could change, Vanda is convinced that governmental structures would need to give more rights to indigenous people such as the Witoto but also to allow them to share their wisdom, expertise, and knowledge, which is so essential in the fight against climate change.

Vanda's long militating efforts to support indigenous people rights paid off in 2019 when she was elected a spokesperson for the Witoto people. This role allowed her to advocate more effectively for her community, and she began to see a slow but significant resurgence of women in leadership roles.

Historically, her people ran under a matriarchal society, and Vanda was determined to restore this balance. She believed that electing more women to leadership positions was crucial for the revival of their traditions and the empowerment of future generations. In 2022, she ran for office, receiving 26,000 votes—an impressive number, though not enough to win the final elections. Nevertheless, the experience strengthened her resolve to continue her fight for justice and recognition.

Vanda has now become a symbol of resilience and hope for her community, leading the charge to reclaim their identity and preserve their rich cultural heritage. "What people do not understand is that when we talk about climate change, we are not talking only about saving the Amazons, but also about saving ourselves" she says.

Her indigenous name means angry ant in the Witotoan language, a fitting title for someone who believes in the strength found in the seemingly insignificant. In her culture, the ant represents community and the power of collective effort to achieve visible results.

Figure 3.1 Vanda Ortega Witoto, an activist and spokesperson of the Witoto tribe.

Endangered ecosystems: Forests facing climate and human pressures

Forests are widely known as the world's green lungs. This connotation derives from their very nature: through photosynthesis, forests use sunlight, water, and carbon dioxide from the atmosphere to produce oxygen (Swingland et al., 2002). They are in fact fundamental for life on the Earth. Thanks to this spontaneous service, forests not only procure their own food to survive but also provide a natural carbon sink, absorbing a net 7.6 billion metric tonnes of CO_2 per year (Harris & Gibbs, 2021). After oceans, forests are the world's largest carbon sinks, covering a third of the Earth's land area (FAO & UNEP, 2020). However, this relationship with carbon is two-sided and damage to the forest's ecosystems comes with its consequences: when trees are cleared or degraded, carbon is then released back into the atmosphere (Harris & Gibbs, 2021). This positions forests at a crucial role with respect to climate change, serving as both a solution and contributing factor (IUCN, 2021b). In fact, while actions of conservation, restoration, sustainable use, and management of forest ecosystems come forward as effective climate mitigation strategies *per se*, the story does not end here, as the ecosystem services provided by forests go well beyond climate mitigation and regulation (FAO & UNEP, 2020).

First and foremost, forests are primary habitats for most of the planet's species. They host a staggering 80% of the world's biodiversity, encompassing the tree and plant species as well as animals and microorganisms that call these ecosystems home (Jenkins & Schaap, 2018). Maintaining and enhancing these lives is crucial to support and increase the resilience of these ecosystems to changing conditions, be it natural or human-induced (Thompson et al., 2009). In fact, the healthier the forests, the stronger their services. In addition, forests ensure and maintain the fertility of soil by releasing nutrients and can promote land preservation especially in the face of a changing climate. Healthy forests also filter out water pollution, recharge aquifers, and reduce the risks related to extreme events and disasters, such as storms and floods.

Second, forests are also crucial for human lives and jobs: globally, 1.6 billion people depend on forests for their livelihoods (IUCN, 2021a). Adding to the incredible value of these ecosystems, forests generate over 86 million green jobs (FAO & UNEP, 2020) and the goods and services they provide to people, such as clean water, healthy soils, food, and medicines, are worth $75–100 billion per year (IUCN, 2021b).

Finally, forests provide an important cultural, spiritual, and recreational value to societies, something that is rooted in indigenous cultures where forests, and the nature they host, are seen as a source of life itself. While indigenous people represent only a small fraction of global population (around 6%) (Beltran & Uchoa, 2024), 70 million of them depend on forests and act as guardians of these ecosystems, playing an incredibly important role in forest protection (Beltran & Uchoa, 2024; UN-REDD, 2022). In fact, some of the most ecologically valuable forests and biodiversity hotspots are found in areas managed by indigenous peoples (Garnett et al., 2018), and their custodianship accounts for 25% of carbon stored by the world's tropical and subtropical forests (Beltran & Uchoa, 2024).

Despite forests being known as a resilient ecosystem, with biodiversity loss, climate change, and pollution on the rise, the provision of all their ecosystem services is at stake and only expected to be increasingly so. Climate, being a key determinant of the state of soil, water, and biodiversity, is a crucial factor in forests' health and functions. While some impacts of a changing climate may be beneficial to forests, several have been found to be detrimental to these ecosystems. A series of natural disturbances are influenced by climate change posing a threat to forests' health, such as storms, droughts and wildfires, invasive alien species, and pests and pathogens (Seidl et al., 2017). All these factors are estimated to already put over 100 million hectares of forests at risk (FAO & UNEP, 2020).

Fires are exacerbated by changing climatic conditions: rising temperatures, longer dry seasons, and droughts make the perfect conditions for forest fires to materialise (Ruiz, 2020). Unfortunately, data shows that this has been the case as these hazards have become more common and widespread and, compared to 20 years ago, tree cover lost through burning has doubled. Due to the relationship between forests and carbon, forest fires create a vicious circle with increasing forest fires in turn increasing carbon emissions, intensifying the impacts of climate change, and leading to an even higher risk of fires in the future (MacCarthy et al., 2024). Similarly, forests' soil is also affected by higher temperatures which lead to the release of carbon dioxide, initiating another loop of negative consequences for tree growth, and their ability to sequester carbon (Ruiz, 2020). Invasive species, pests, and diseases have also become more common, representing another direct threat to forest health, sustainability and productivity. It has been found that 35 million hectares of forests are adversely affected by forest insect pests on a yearly basis (FAO, 2010).

Alongside these impacts, forests are being cleared intentionally, further leading to their deterioration. Deforestation and forest degradation are in fact the biggest threats faced by forests. When degraded, forests reach a state in which they are no longer able to deliver their services and provide goods (IUCN, 2021a). Accounting for differences across regions, deforestation is driven by several human activities, and primarily by agriculture, including both cropland expansion and livestock grazing, which together contribute to almost 90% of global deforestation (FAO, 2022). Other drivers include urban and infrastructure development, extractive industries, such as energy production, timber logging, and mining, and factors such as the above-mentioned wildfires (Pacheco et al., 2021). In addition, the impacts of deforestation can be felt well beyond where it happens, through trade. The conversion of forests to other land uses has caused the loss of 420 million hectares of forests since 1990. In the same period, the global forest area has decreased by 178 million hectares. While the rate of deforestation has slowed down over the past decades, it continues to represent a threat, together with climate change. Since the 1960s, more than half of the tropical forests have been destroyed (FAO & UNEP, 2020). As a direct result of losing these ecosystems, biodiversity is losing its home and in some cases it is changing drastically, with livelihoods being profoundly impaired.

As forests become increasingly affected by climate change and intensive human activities, they may reach a point of no return. Climate models forecast that more than half of the Amazon, the world's largest rainforest, will turn into savanna in the next 50 years. If an ecological tipping point is reached, it will become impossible for the environment to support this ecosystem further (Hanbury, 2020).

As is often the case, the impacts of climate change and of deforestation are felt differently across communities. Those more reliant on forests for their income and subsistence may find themselves in the position of having to move away to sustain their livelihoods. At the same time, their health is put at risk due to worsening air and water conditions and through the loss of forest ecosystem services (Shyamsundar et al., 2021). Because of these ripple impacts, people may be forced into poverty. With one million people depending on forests living in poverty already, the situation is particularly dire (IUCN, 2021a). Indigenous forest communities are continuously finding themselves in this position, facing the risk of losing their traditional land, and with it their own identity. In this way, forests are losing their best custodians (Kanungo, 2023).

All of this is ultimately the result of a deep change in the human-forest relationship, which has shifted from intrinsic values to mere economics. Traditional knowledge based on spiritual and deep relationships between human and nature is at risk of being lost. Bringing back these values is about acknowledging the intangible contributions of forests and their biodiversity to people, reflecting on the key role that forests can play in providing the opportunities to lift people out of poverty. For this, a balance should be found between ensuring forest conservation and meeting local needs and the demand for resources, something which requires biodiversity to be mainstreamed in all forest management practices (FAO & UNEP, 2020).

Bibliography

Beltran, L., & Uchoa, P. (2024). Lessons from Indigenous Leaders to Protect the Amazon Rainforest. Available from: https://www.weforum.org/agenda/2024/01/lessons-from-indigenous-leaders-to-protect-the-amazon-rainforest/ [Last accessed: 15th January 2025].

FAO (2010). *Global Forest Resources Assessment 2010*. Food and Agriculture Organization of the United Nations. Available from: https://www.fao.org/4/i1757e/i1757e00.htm [Last accessed: 15th January 2025].

FAO (2022). *FRA 2020 Remote Sensing Survey* (FAO Forestry Paper Issue). Food and Agriculture Organization of the United Nations.

FAO & UNEP (2020). *The State of the World's Forests 2020. Forests, Biodiversity and People*. Food and Agricultural Organization of the United Nations and United Nations Environment Program.

Garnett, S. T., Burgess, N. D., Fa, J. E., Fernández-Llamazares, Á, Molnár, Z., Robinson, C. J., Watson, J. E. M., Zander, K. K., Austin, B., Brondizio, E. S., Collier, N. F., Duncan, T., Ellis, E., Geyle, H., Jackson, M. V., Jonas, H., Malmer, P., McGowan, B., Sivongxay, A., & Leiper, I. (2018). A Spatial Overview of the Global Importance of Indigenous Lands for Conservation. *Nature Sustainability*, 1(7), 369–374. https://doi.org/10.1038/s41893-018-0100-6.

Hanbury, S. (2020). Impending Amazon tipping point puts biome and world at risk, scientists warn. Available from: https://news.mongabay.com/2020/01/impending-amazon-tipping-point-puts-biome-and-world-at-risk-scientists-warn/ [Last accessed: 15th January 2025].

Harris, N., & Gibbs, D. (2021). *Forests Absorb Twice As Much Carbon As They Emit Each Year*. World Resource Institute. Available from: https://www.wri.org/insights/forests-absorb-twice-much-carbon-they-emit-each-year# [Last accessed: 22nd September 2024].

IUCN (2021a). *Deforestation and Forest Degradation*. International Union for Conservation of Nature. Available from: https://iucn.org/resources/issues-brief/deforestation-and-forest-degradation [Last accessed: 15th January 2025].

IUCN (2021b). *Forests and Climate Change*. International Unit for Conservation of Nature. Available from: https://iucn.org/resources/issues-brief/forests-and-climate-change [Last accessed: 15th January 2025].

Jenkins, M., & Schaap, B. (2018). *Forest Ecosystem Services - Background Study Prepared for the Thirteenth Session of the United Nations Forum on Forests*. United Nations Forum on Forests Global Forest Goals. Available from: https://www.un.org/esa/forests/wp-content/uploads/2018/05/UNFF13_BkgdStudy_ForestsEcoServices.pdf [Last accessed: 15th January 2025].

Kanungo, A. (2023). *The Silent Cry of the Forest: How Deforestation Impacts Indigenous Communities*. Available from: https://earth.org/the-silent-cry-of-the-forest-how-deforestation-impacts-indigenous-communities/ [Last accessed: 15th January 2025].

MacCarthy, J., Richter, J., Tyukavina, S., Weisse, M., & Harris, N. (2024). *The Latest Data Confirms: Forest Fires Are Getting Worse*. World Resources Institute. Available from: https://www.wri.org/insights/global-trends-forest-fires [Last accessed: 20th January 2025].

Pacheco, P., Mo, K., Dudley, N., Shapiro, A., Aguilar-Amuchastegui, N., Ling, P. Y., Anderson, C., & Marx, A. (2021). *Deforestation Fronts: Drivers and Responses in a Changing World*. WEF. Available from: https://wwfint.awsassets.panda.org/downloads/deforestation_fronts___drivers_and_responses_in_a_changing_world___full_report_1.pdf [Last accessed: 20th January 2025].

Ruiz, S. (2020). *Climate Change Will Alter Forests as We Know Them*. Global Forest Watch. Available from: https://www.globalforestwatch.org/blog/forest-insights/climate-change-will-alter-forests-as-we-know-them/ [Last accessed: 15th January 2025].

Seidl, R., Thom, D., Kautz, M., Martin-Benito, D., Peltoniemi, M., Vacchiano, G., Wild, J., Ascoli, D., Petr, M., Honkaniemi, J., Lexer, M. J., Trotsiuk, V., Mairota, P., Svoboda, M., Fabrika, M., Nagel, T. A., & Reyer, C. P. O. (2017). Forest Disturbances under Climate Change. *Nature Climate Change*, 7, 395–402. https://doi.org/10.1038/nclimate3303.

Shyamsundar, P., Sauls, L. A., Cheek, J. Z., Sullivan-Wiley, K., Erbaugh, J. T., & Krishnapriya, P. P. (2021). Global Forces of Change: Implications for Forest-Poverty Dynamics. *Forest Policy and Economics*, 133, 102607. https://doi.org/10.1016/j.forpol.2021.102607.

Swingland, I. R., Bettelheim, E. C., Grace, J., Prance, G. T., Saunders, L. S., Malhi, Y., Meir, P., & Brown, S. (2002). Forests, Carbon and Global Climate. *Philosophical Transactions of the Royal Society of London. Series A: Mathematical, Physical and Engineering Sciences*, 360(1797), 1567–1591. https://doi.org/10.1098/rsta.2002.1020.

Thompson, I., Mackey, B., McNulty, S., & Mosseler, A. (2009). *Forest Resilience, Biodiversity, and Climate Change. A Synthesis of the biodiversity/resilience/stability Relationship in Forest Ecosystems*. (Technical Series, Issue). Secretariat of the Convention on Biological Diversity. Available from: https://www.cbd.int/doc/publications/cbd-ts-43-en.pdf [Last accessed: 20th January 2025].

UN-REDD (2022). *Recognizing and Empowering Indigenous Peoples and Local Communities as Critical Partners in Forest Solutions to the Climate Emergency*. UNDP, FAO. Available from: https://www.climateandforests-undp.org/sites/default/files/resources/Final%20IP%20Brief.pdf [Last accessed: 15th January 2025].

4 Mangroves guardians

As a cheerful rooster resonates in Santiago's courtyard, his eyes seem to get tinged with the deep, misty colours of the surrounding forest while he dreams about his youth. The 80-year-old hints a smile and mutters: "My father was a fisherman. And I still remember the sense of excitement I had in the early mornings when he'd take me out to catch crabs in the mangrove forest. This is why I decided to come back here today". Green, brown, and grey, and all the shades of mangroves in between, painted the background of his childhood.

Santiago is now heading a fisherman's organisation in Puerto El Morro, in the province of Guayaquil, dedicated to fishing mostly crabs and shellfish, all species that thrive in the mangrove forests across the Gulf of Guayalquil's coastline.

Mangroves in Ecuador make up more than half of the country's Pacific coastline, Santiago claims, and serve as a refuge to shellfish and young fish, providing them with a nurturing environment until they grow sufficiently to venture into the open waters. But their role does not end here: mangroves are a natural coastal protection from flooding and erosion, and their maintenance is therefore crucial for coastal communities like Santiago's. The beating heart of mangroves in the country is exactly the Gulf, where more than 80% of mangroves are located.

Today, Santiago knows that he is exactly where he should be, in a place that has always been his home, protecting this delicate ecosystem from the numerous threats it has faced throughout the years. But it wasn't always like this: when he was young, he had wanted to become a teacher and decided to move away from Puerto El Morro to pursue his dreams. During his visits back to the small town from the largest city nearby, he witnessed how the community of fishermen, including his father, were struggling to survive. This was due to the progressive use of the mangroves ponds for shrimps rather than crabs, spurred by the shrimp farming industry.

While deforestation had taken place in the area already since the 1940s, Santiago explains how this "second wave of destruction" happened around 40 years ago.

> I remember seeing from one day to the other large bulldozers, driven by shrimp farmers, clearing swathes of the mangroves to make way for their ponds. They saw only profit, but I knew they were destroying something

DOI: 10.4324/9781003493662-5

> sacred, something that had been there long before them, or any of us, and on which our community still depends.

It was around 20 years ago, when Santiago met Raúl Carvajal, first colleague and later a friend, who was involved in the work of the NGO *Conservación Internacional de Ecuador*. Raúl was aware of how the shrimp industry, coupled with the impact of a changing climate, was already largely destroying the Gulf and therefore started a movement to push for a ministerial agreement to declare mangrove-rich areas as protected areas and devolve property rights over the mangroves to ancestral communities. The idea behind creating these areas was to transform the communities' mere fishermen into "mangroves guardians" to ensure that they would oversee sustainable fishing practices and monitor the mangrove deforestation, enabling this rich ecosystem to thrive and continue to deliver its many functions.

Besides the challenge of deforestation, Santiago is also aware of how climate change is making things more challenging.

> In the last few years, I realised how it's not easy to foresee how and when El Niño will arrive. It used to be December, but now it's always either earlier or later. At the beginning, the ocean gets warmer, and this is good for fish and shellfish. But El Niño also brings a lot of rain and high tides, which make sea water sweeter and weakens mangrove ecosystems, killing shellfish. It also makes crabs run away for saltier waters, so we do not know where they are and cannot capture them.

Despite approaching retirement age, Santiago knew he was not going to sit still while his community and family were facing such drastic challenges. "I knew I had to protect this land from the increasing natural hazards, and to stop mangroves from being cut". He immediately decided to follow the negotiations on the protected areas and worked alongside Raúl to enlarge the concessions areas across the Gulf. That's when he picked up his role of "mangrove guardian", heading a small organisation of 36 fisherwomen and fishermen involved in sustainable fishing and mangrove protection.

"These laws, defining the protected areas for our land, have been a huge success for our communities. But it's not easy to implement them", Santiago continues.

> We structured ourselves across small organisations of fisherwomen and fishermen committed to sustainable practices. And these practices are very strict as we have to abide to the rules of biodiversity: for instance, we have specific months of the year where we cannot fish red or blue crabs, as they are either in the phase of moulting and cannot be eaten, or in the phase of mating and they are very vulnerable.

These strict practices allow to maintain biodiversity in the area and lower the risks tied to the increasing challenges from El Niño.

> We also get instructions about how large crabs and seashells should be when fished. This means a lot of work to make sure that our communities are respecting these standards. It is exactly what I am doing everyday: I wake up around three in the morning, and, after a few chores around the house, I do rounds of checks in small boats to ensure no deforestation is happening. Phones don't work in most of the protected areas, so we use a Motorola radio to communicate, and some colleagues also have a drone which we use to take pictures to share with the government.

Santiago also manages weekly assemblies, where his role is to report on the impact of their fishing practices. "Not everyone embraces this new

Figure 4.1 Santiago Morales, head of a fishermen's organisation, in front of a mangrove forest.

concept completely – and some fishermen want to stay independent and not abide by the rules". He needs to report when this is happening and ensure that fines or other measures are put in place. "Rules might not be pleasant in the beginning" he claims,

> but rules are what saved us here, in this community. Thanks to the government's desire to maintain this delicate and rich ecosystem, we were able to avoid a massive deforestation wave, which would have risked destroying Ecuador's fishing practices and cultural heritage. That is why I am proud of giving this honest service to the community around me.

At 80 years of life, Santiago has seen a whole ecosystem being threatened in front of his eyes. Despite the mounting challenges, he is not sitting on his hands: "I have not lost hope or else I would not be here at this age" he smiles. "I am here to protect this ecosystem everyday, until the last day of my life, and I hope others will come after me to do the same".

Mangrove ecosystems sustaining livelihoods

One might say that mangroves are the breathing lungs of coastal ecosystems. On the bottom, stretches of thick exposed supporting roots, known as "breathing roots", branch out smoothly in the adjacent waters, taking in oxygen from the air and releasing into the ground. On the top, green and waxy foliage can reduce the amount of water lost through evaporation, acting like a natural water-saving umbrella (Kathiresan & Bingham, 2001). But these are not just any peculiar trees, they come with a clear mission: normally located in intertidal zones, they provide a buffer between terrestrial and marine and freshwater ecosystems, protecting the communities that call these areas home. They act as nature defenders, preventing shorelines' erosion and flooding, providing food security and storing carbon. At least 210 million people around the globe live within 10 km of mangroves, depending on these ecosystems for their livelihoods through mangrove fisheries (Hutchison et al., 2014).

Mangroves create the ecosystem for life on earth to thrive. As rich and productive habitats, mangroves are both a nursery and a feeding ground for fish, shrimps, crabs, insects, and birds, and provide valuable resources including timber, wild forest products, and clean water. More than 1,500 plant and animal species are found to depend on mangroves (UNEP, 2023) and almost 80% of global fish catches are directly or indirectly relying on these ecosystems. And their support is not only limited to animals, but they are also crucial for humans: indirectly, by supporting fisheries, they bring an attractive source of tourism, and through restoration projects, mangroves also provide important employment opportunities to local communities (IUCN, 2021).

In addition to nurturing flourishing ecosystems, mangroves play a pivotal role in safeguarding our natural world. Mangroves act as a natural shield against natural hazards, such as storms, tsunamis, and coastal erosion, as well as sea level rise. By thriving in coastal settings, they have developed a natural ability both to cope with such hazards and to attenuate their impacts (Spalding et al., 2014). As an example of their heroic functions, mangrove forests can reduce wave heights by 31%, therefore reducing flood risks to homes, property, and infrastructure (IUCN, 2021). Similarly, the damaging impacts of hurricanes are found to decline with the presence of wide mangrove belts, stressing the need for large scale mangrove restoration (del Valle et al., 2020). As a matter of fact, restoring forests of mangroves has proved to be an effective nature-based solution, providing clear benefits to biodiversity and people while addressing societal challenges. In addition, mangrove forests are incredibly effective carbon sinks. If left undisturbed, these ecosystems can absorb large amounts of carbon dioxide and store three to four times more carbon than terrestrial forests (Nyanga, 2020), locking it for hundreds to thousands of years in the soil beneath them (IUCN, 2021). Safeguarding these ecosystems, for instance, through protected areas, will also protect the ecosystem services they provide and the biodiversity they host, while reducing the risks deriving from climate hazards (IUCN, 2019).

Coastal mangroves have an immense value to their local economy, estimated to be between $33,000 and $57,000 per hectare annually for developing countries hosting these ecosystems (Hernández-Blanco et al., 2021; The Commonwealth, 2024). But the real value of mangroves, as it is the case with many natural resources, is often only appreciated when these habitats are lost. Now more than ever, mangroves are under increasing threats and are disappearing at an alarming rate. Deforestation, climate change, and coastal development represent only some of the biggest threats to these ecosystems.

Mangroves make up a very tiny fraction of our planet's land, just 0.7%. However, mangrove deforestation accounts for up to a staggering 10% of global deforestation emissions (Mulhern, 2020). These forests are falling victims of conversion of the land they occupy to aquaculture, agriculture, and urban land uses, all leading to the loss of these forests, a transformation hardest in developing countries where 90% of these species are found (Morocho et al., 2022). Historically, the expansion of shrimp farming has resulted in significant mangrove deforestation. This has been particularly the case in countries such as Ecuador where more than 70% of mangrove ecosystems were destroyed by shrimp ponds (Benavides, 2021). Even though the industry has slowed down its expansion and a positive balance in mangrove recovery can now be observed, deforestation consequences are still representing the main driver for mangroves' alteration and loss (Lopez, 2018). Policies concerning property rights have been explored in the context of forests as a means to reduce deforestation. In Ecuador, rights assigned to indigenous people and local communities (IPLCs) in mangrove ecosystems have proved successful in this effort, while recognising the role played by the institutional and market settings in place (Tanner & Ratzke, 2022). Other issues still present are illegal land concessions and authorisations for mangrove clearing failing to meet the legally established requirements (Morocho et al., 2022; Romero, 2014).

In this already gloomy picture, climate change has been impacting mangrove ecosystems, and continues to do so, with more frequent and intense storms and heavy rainfall events, increased coastal erosion and temperatures, sea level rise, and changing ocean currents (Ward et al., 2016). Increased coastal flooding due to sea level rise is harming mangroves because of their sensitivity to changes in inundation duration, frequency, and salinity levels, leading to reduced ecosystem services and productivity. Similarly, the occurrence of storms such as hurricanes and cyclones can damage mangroves through wave activity and wind damage. Changing rainfall patterns are also likely to influence mangroves' distribution and growth rates due to changes in soil and water salinity. Further, temperature plays a key role in mangroves' assimilation of CO_2, and increasing temperatures are found to reduce their productivity (Ward et al., 2016).

According to climate scenarios, all these factors are expected to worsen in the future. There comes a paradox: climate change is posing yet increasing threats to mangrove ecosystems, while at the same time accentuating their vital role in reducing the vulnerability of their adjacent coastal areas. With the current climate and mangrove coverage, we see around 3.5 million people and GDP worth $400 million being at risk due to loss of coastal protection from mangroves.[1] However, the expected impacts of climate change[2] will double and increase nearly threefold the population and GDP at risk, respectively (Blankespoor et al., 2017).

These findings paint a dire picture but are also a clear call to action to ensure that these important ecosystems are protected, conserved, and restored to the extent possible, enabling their vital functions and ensuring they meet their noble mission: enabling and protecting natural livelihoods.

Notes

1 Based on developing countries in four regions – East Asia-Pacific, South Asia, Africa, and Latin America and Caribbean – and in only in those countries with previous exposure to tropical cyclones (representing more than 50% of global mangroves).
2 Future climate scenario out to 2,100 with a 1-metre sea level rise and 10% intensification of storms.

Bibliography

Benavides, M. T. (2021). *Conflictos en el ecosistema manglar de la costa del Ecuador. el desarrollo de la acuacultura industrial del camarón frente a los derechos de los pueblos de recolectores y pescadores de los estuarios Periodo: 2008 – 2019 [Conflicts in the Mangrove Ecosystem on the Coast of Ecuador. The Development of Industrial Shrimp Aquaculture Versus the Rights of the Gatherers and Fishermen of the Estuaries.*

Period: 2008 – 2019]. Corporación Coordinadora Nacional para la Defensea del Ecosistema Manglar [National Coordinating Corporation for the Defense of the Mangrove Ecosystem]. Available from: https://ccondem.org.ec/wp-content/uploads/2021/11/D-Conflictos-en-el-ecosistema-manglar-de-la-costa-del-Ecuador-1.pdf [Last accessed: 10th September 2024].

Blankespoor, B., Dasgupta, S., & Lange, G.-M. (2017). Mangroves as a Protection from Storm Surges in a Changing Climate. *Ambio*, 46(4), 478–491. https://doi.org/10.1007/s13280-016-0838-x.

del Valle, A., Eriksson, M., Ishizawa, O. A., & Miranda, J. J. (2020). Mangroves Protect Coastal Economic Activity from Hurricanes. *Proceedings of the National Academy of Sciences*, 117(1), 265–270. https://doi.org/doi:10.1073/pnas.1911617116.

Hernández-Blanco, M., Costanza, R., & Cifuentes-Jara, M. (2021). Economic Valuation of the Ecosystem Services Provided by the Mangroves of the Gulf of Nicoya Using a Hybrid Methodology. *Ecosystem Services*, 49, 101258. https://doi.org/10.1016/j.ecoser.2021.101258.

Hutchison, J., Spalding, M., & zu Ermgassen, P. (2014). *The Role of Mangroves in Fisheries Enhancement*. The Nature Conservancy and Wetlands International. Available from: https://www.oieau.fr/eaudoc/system/files/33226.pdf [Last accessed: 10th September 2024].

IUCN (2019). *Protected Areas and Climate Change*. International Union for Conservation of Nature. Available from: https://www.iucn.org/sites/default/files/2022-07/protected_areas_and_climate_change_briefing_paper_december_2019-final.pdf [Last accessed: 10th September 2024].

IUCN (2021). *My Mangroves, My Livelihood*. International Union for Conservation of Nature. Available from: https://www.iucn.org/news/oceania/202107/my-mangroves-my-livelihood [Last accessed: 10th September 2024].

Kathiresan, K., & Bingham, B. L. (2001). Biology of Mangroves and Mangrove Ecosystems. In: A. J. Southward, C. M. Young, L. A. Fuiman, & P. A. Tyler (Eds.) *Advances in Marine Biology*. Academic Press, pp. 81–251. https://doi.org/10.1016/S0065-2881(01)40003-4.

Lopez, F. (2018). Mangrove Concessions: An Innovative Strategy for Community Mangrove Conservation in Ecuador. In: C. Makowski, & C. Finkl (Eds.) *Threats to Mangrove Forests: Hazards, Vulnerability, and Management*. Springer Cham, pp. 557–578. https://doi.org/10.1007/978-3-319-73016-5_25.

Morocho, R., González, I., Ferreira, T. O., & Otero, X. L. (2022). Mangrove Forests in Ecuador: A Two-Decade Analysis. *Forests*, 13(5), 656. https://www.mdpi.com/1999-4907/13/5/656

Mulhern, O. (2020). *The True Value of Mangroves*. Available from: https://earth.org/data_visualization/the-true-value-of-mangroves/ [Last accessed: 25th June 2024].

Nyanga, C. (2020). The Role of Mangroves Forests in Decarbonizing the Atmosphere. In: B. Mattia, F. Marco, & R. Luca (Eds.) *Carbon-Based Material for Environmental Protection and Remediation*. IntechOpen, pp. 153–164. https://doi.org/10.5772/intechopen.92249.

Romero, N. S. (2014). Neoliberalismo e Industria Camaronera En Ecuador [Neoliberalism and Shrimp Industry in Ecuador. *Letras Verdes. Revista Latinoamericana De Estudios Socioambientales FLACSO - Ecuador*, 15, 55–78.

Spalding, M., McIvor, A., Tonneijck, F. H., Tol, S., & van Eijk, P. (2014). *Mangroves for Coastal Defence. Guidelines for Coastal Managers & Policy Makers*. Wetlands International and the Nature Conservancy. Available from: https://www.nature.org/

media/oceansandcoasts/mangroves-for-coastal-defence.pdf [Last accessed: 10th September 2024].

Tanner, M., & Ratzke, L. (2022). Deforestation, Institutions, and Property Rights: Evidence from Land Titling to Indigenous Peoples and Local Communities in Ecuador. *Corporación Andina de Fomento (CAF) - Working Paper*. Available from: https://scioteca.caf.com/handle/123456789/1995 [Last accessed: 10th September 2024].

The Commonwealth (2024). *Mangrove Ecosystems and Livelihoods*. Available from: https://thecommonwealth.org/bluecharter/mangrove-restoration [Last accessed: 25th June 2024].

UNEP (2023). *Decades of Mangrove Forest Change: What Does It Mean for Nature, People and the Climate?* United Nations Environment Programme. Available from: https://www.unep.org/resources/report/decades-mangrove-forest-change-what-does-it-mean-nature-people-and-climate [Last accessed: 10th September 2024].

Ward, R. D., Friess, D. A., Day, R. H., & Mackenzie, R. A. (2016). Impacts of Climate Change on Mangrove Ecosystems: a Region by Region Overview. *Ecosystem Health and Sustainability*, 2(4), e01211. https://doi.org/doi:10.1002/ehs2.1211.

5 The nostalgia of the sea

> You know what we used to say? El mar es una abuela - the sea is like a grandma. Throughout my whole life, the sea was always around me, everywhere - it was so present, so much part of our everyday life, that I could not think about anything else, even in my dreams.

Evelio Lopez Morales is a primary school teacher from the Guna community, indigenous people of Panama and Colombia. He was born in the main island-capital of the Guna-Yala archipelago, off the Panamanian coast, called Gardi Sugdub. For all his life, up until recently, he commuted every week between his city-island and a nearby community, Soledad Neria, where he taught in a primary school. For him, traveling in a small boat was as normal as it is for many in big cities to travel by bus, car, or bike – with the exception that, instead of morning city traffic around him, his trip was accompanied by the sound of the waves and the blue of the sea.

However, the island of Gardi Sugdub recently underwent a massive relocation project, with over 90% of its population being moved to the mainland. While his job did not change, his commute did, as he now lives between the new mainland site, where he spends the weekends, and the island of Soledad Neria, where he teaches.

"This might seem like a big change - and it was, but it also was nothing new for us. The truth is that our community started thinking of moving already more than 15 years ago", Evelio tells us. The concept of relocation was already rooted in Gardi Sugdub's inhabitants when the government, acknowledging the pressure that the city-island was facing due to climate change and overpopulation, decided to look into an alternative site.

> I still remember the day we decided to move. It was a humid November evening, and I was on the patio of a small house near the shore, talking with some Spanish friends who had arrived earlier that day on a sailing boat. We shared stories and laughter over glasses of wine. All of the sudden, as the night deepened, the sea began to swell, creeping higher onto

DOI: 10.4324/9781003493662-6

> the shore than usual, while the nice breeze became a steady wind from the Northeast. By morning, the rising waters and the strong winds had damaged many houses on the island.
>
> On the same day, the community was urgently convened in the Casa Grande, our spiritual and decision making center. That night, the Saila, a spiritual leader, sang the ancient songs of their people and the Argal, the interpreter, explained the deeper meaning behind his chants: the Guna, who had traditionally lived near the forest, needed to return to the mainland. It was then when our community as a whole took the decision to look into a project of moving out from the island.

This decision, Evelio tells us, did not come out of the blue: at the time, the island was already changing a lot, as coral reefs were slowly disappearing and there were less and less fish in the surrounding areas. As such, Gardi Sugdub inhabitants had reinvented themselves, and turned the island into a touristic destination.

> Once we opened up to the western world, the subjectivity of people changed. But this was also a great opportunity for many, and people from other islands started coming to Gardi Sugdub in search of a job. This included people who had been living in Panama's main cities and who moved to the island looking for permanent or temporary residence. Slowly, but very steadily, the island was becoming overpopulated.

The increasing frequency of violent storms, coupled with the growth in population, made the fragility of the island more and more visible, and its people could not help but to notice this change. Entire homes were frequently swept away from the tides, the shores were slowly eroding, and the rising of the sea was making living conditions very hard. "Even if the Gunas might not be fully aware of climate change, nor of the science behind it," Evelio says

> we all noticed the sea level rising on many of our islands, not only on Gardi Sugdub. We were affected firsthand by the consequences of very violent storms, and we understood the need to look for a better future on the shore. However, the process of moving to the mainland turned out to be much harder than what we had anticipated.

The relocation process was finally realised only in June 2024, more than a decade after initial discussions took place. A new mainland site was built by the government, with a school, a hospital, and houses for each of the families. However, not all families agreed with the relocation: especially for older people and people who were so used to living surrounded by the sea, it was hard to think of moving to such a distant and, in a way, abstract place.

"To date, while around 300 families have now fully moved to the new site, 30 or 40 still continue living on the island", Evelio tells us. For him, the process was bumpy but overall positive, as he acknowledges how it originated from a need of the community and was not an imposition of the government. Still, during the process, the Guna had to strive to achieve the delicate balance between tradition and novelty. Unavoidably, they all knew that something was going to be lost in the process.

"The move happened very recently, just a few months ago. And there are still a lot of problems with the new site" continues Evelio.

> For instance, the construction of the homes did not consider important details of our culture, as we traditionally sleep in hammocks, and we were provided with Western beds. For us, these look very uncomfortable. Also, the water pressure for drinking water still remains quite insufficient. And, at the moment, we do not have a proper waste management system.

Evelio is aware that Guna's life will not be the same in the new site. A certain degree of adaptability will make it so that the mainland neighbourhood will be more influenced by local customs and tradition – for instance, a Catholic church has been built around the site, which was accepted by the community but which will deeply influence Guna spirituality.

However, for Evelio, some traditional practices cannot be forgotten in the new neighbourhood. Both the Casa Grande, which functions as a House of Congress, and the Innanega, the house where ceremonies for girls entering into puberty are performed, are key to the existence and survival of the Guna. This is why the two houses were a first prerequisite to convince Guna families to move from Gardi Sugdub.

In this new life, Evelio still commutes during the week as he travels by boat from the mainland to the island of Soledad Neria, where he stays for five days to teach at the primary school. On the weekend, he then goes back to the mainland. When asked about whether other Guna islands are also thinking about relocation, he says

> this depends on how Gardi Sugdub's project is going to turn out. Some communities are also aware of how fragile life on these small islands might become in the years to come, and are looking with curiosity at the new mainland site. But some communities are also quite against this. It is not easy to leave one's life, and land, behind.
>
> In many ways, I think our quality of life in the new community will improve. We will not be as crowded, and have access to more services. But change, even when positive, always leaves space for fear of the unexpected. I sometimes worry about how Guna's life is going to change. One thing is sure: we will miss the sea surrounding us.

Figure 5.1 Evelio Lopez Morales, part of the Guna community and primary teacher, in front of a house.

"My dreams have been inspiring the stories I write". Even there, explains Evelio, the sea is always present, not only as the setting of my stories but as the protagonist as well. One story, born in the quiet of Soledad Neria, was about a boy and a dolphin. In the dream, the dolphin took the boy on a journey, showing him a place in the sea that was full of fish. When the boy told his father about the dream, the father trusted him to embark on a small journey, and they found the spot just as the magic dolphin had shown.

In many ways, the story mirrored his own life. The sea had always guided him, even now as it rose and pushed his community to new shores. The Guna people had already survived so much –division, repression, and revolution. Evelio believes they would survive this too, as long as they stayed true to their vision and values.

The unavoidable consequences of sea level rise

As such an essential part of life, and a crucial resource for humanity, the sea has always represented one of those stable truths about life on Earth. Oceans provide half of the oxygen we breathe: they are, by definition, sustaining us (National Ocean Service, n.d.). The sea has also traditionally been an important cultural element, conceptualised across various cultures and reflecting the contradictory ways in which it has been experienced. A symbol of power, yet danger, of calmness and stability, yet chaos, represented over time in art, literature, and mythology. In current times, while seas and oceans still represent a vital aspect of life, some threats to their stability have emerged in new forms. Sea level rise, referring to the average increase in volume and height of oceans' surface, is one of them. While this phenomenon has materialised already throughout the history of life on Earth, it has now reached unprecedented rates during the past 3,000 years, becoming a crucial indicator of global change and human-induced climate change (Gulev et al., 2021; NASA, n.d.; UN Secretary-General's Climate Action Team, 2024).

Sea level rise is the direct consequence of two main factors: the melting of ice sheets and glaciers increasing the volume of water, and the rising temperature of the sea. As the sea warms, seawater expands, adding volume to seas and oceans (NASA, n.d.). Both factors are generated by climate variability, including anthropogenic climate change, making sea-level rise a direct symptom of such change (WCRP Global Sea Level Budget Group, 2018). In particular, since 1970, anthropogenic climate change has become the main driver of sea level rise (Oppenheimer et al., 2019). As temperature rises due to growing emissions, glaciers and ice sheets melt, the temperature of the water also increases, further contributing to rising water levels (UN Secretary-General's Climate Action Team, 2024). Since 1993, the global mean sea water level has risen by 10.3 centimetres (CMEMS, 2024). But the biggest shock came in 2022, a year which marked the highest sea level increase ever recorded (CMEMS, 2024). Looking ahead, the outlook from climate scientists is deeply concerning. Projections indicate an even higher increase in sea level under all future scenarios (Oppenheimer et al., 2019). The rates of ice loss from the Greenland and Antarctic ice sheet are increasing, exacerbating the crisis. Alarmingly, the seven worst years on record have all been occurring in the last decade (Otosaka et al., 2023). Considering a fast disintegration of the Antarctic ice sheet – a scenario, however, fraught with deep uncertainty – global mean sea level rise is expected to reach even 2.3 metres by the end of the century, and 5.4 metres by 2150 (EEA, 2025; Oppenheimer et al., 2019).

While many people around the globe still ignore these scenarios, many others are already seeing and living the consequences of such change and the threat of sea level rise has already become a reality for them. Already now, more than 250 million people globally live on land less than 2 metres above sea level. Considering a 1-metre sea level rise and no population growth, this number could reach 410 million people by 2100. Of these, over 70% are expected to be in the tropics, which is already the area most at risk (Hooijer & Vernimmen, 2021).

Islands are also unsurprisingly being hit the hardest from sea level rise. For instance, in Small Island Development States (SIDS), hosting 70 million people, the impacts of climate change have long been threatening their coastlines and livelihoods. Already in 1990, these were recognised as one the most vulnerable regions to climate change by the Intergovernmental Panel on Climate Change. Despite their geographic and environmental differences, all SIDS share a common challenge: rising sea levels and intensifying cyclones and storms are driving up flooding risks (Vousdoukas et al., 2023). In the Pacific region, islands such as Vanuatu and Kiribati – rising just a few metres above sea level – are facing the highest rates of sea level rise (Freestone & Çiçek, 2021). The future looks even more precarious, as risks of coastal flooding are in fact expected to increase further even under stringent mitigation scenarios. Already, the annual costs of flood-related damages for SIDS have reached $1.65 billion (Vousdoukas et al., 2023).

In the future, almost all coastal areas around the world (90%) will be affected to some extent (WEF, 2019): by 2050, it is estimated that 800 million people will live in cities where sea level will rise by more than half a meter (UCCRN, 2018). Coastal areas, historically favoured for human settlements, have become hotspots of vulnerability. Not only do they contend with rising sea level, but are also grappling with other pressures, such as population growth, urban development, and coastal erosion (Reimann et al., 2023).

While coastal populations continue to grow, so do the risks they are bound to face (Reimann et al., 2023). These zones are home to vital economic activities, critical infrastructure as well as cultural heritage sites (Marzeion & Levermann, 2014). In fact, 14% of the global GDP is generated in low-elevation coastal zones and rising no more than 10 metres above it (Magnan et al., 2022). This makes the impacts of hazards in coastal areas significant and socio-economic dynamics play an important role in determining exposure (Reimann et al., 2023). People's dependence on the sea is not only demonstrated by how populous coastal areas are but also by the fact that many economic activities are located in proximity of the sea (Magnan et al., 2022).

Both slowly retreating coastlines and fast floods generated by sea level rise are already reshaping our geographies. However, the impacts of this change go well beyond the direct damages from these events, negatively affecting entire livelihoods and cultures (Freestone & Çiçek, 2021). Reduced or even loss of economic activities – such as agriculture, fisheries, and port activities – resulting from sea level rise impacts, may damage global food systems and supply chains, and potentially lead to negative ramifications in our economy and geo-politics (Glavovic et al., 2022; UN Secretary-General's Climate Action Team, 2024). Similarly, at local levels, traditional diets, recreational activities as well as aesthetic aspects might change drastically, threatening cultural dimensions and putting at risk local and indigenous knowledge (Freestone & Çiçek, 2021).

Practical infrastructural solutions to the challenge of sea level rise are being explored. On the one hand, the reliance on grey infrastructure and geo-engineering is seeing advances in the development of sea walls and similar hard barriers. Yet, it is nature itself that is often providing innovative solutions. Nature-based solutions, such as restored mangroves forests, corals, saltmarshes, and seagrass meadows, are offering a natural shield and coastal protection against erosion and sea level rise. Hundreds of millions of people worldwide are already being protected this way against storm surges and waves (Magnan et al., 2022).

As the impacts of climate change and sea level rise become more catastrophic, we are left to rely on all possible solutions. Among the most profound consequences is the growing wave of climate-induced migration. Around the world, communities are already being forced to move to inland areas as rising waters threaten physical land and local economies are put at risk (Glavovic et al., 2022; Causse et al., 2016).

In many cases, when infrastructure solutions are not applicable, governments have responded with relocation strategies, moving entire communities to a new settlement. These strategies go beyond the simple shifting of people away from risks, as they are aiming to reinvent and regenerate lives and livelihoods (Causse et al., 2016). Yet, this shift is anything but simple. Leaving one's life behind comes with strong emotional repercussions and potentially threatens the loss of local memory, culture, and traditions (Ietto et al., 2022).

Still, displaced individuals demonstrate remarkable resilience. Behind the courage of starting anew lies the challenge and vulnerability of the gesture. The experience will likely remain deeply rooted in them, even after starting and establishing a new life elsewhere, further away from the rising waters.

Bibliography

Causse, C., Mokhnacheva, D., & Camus, G. (2016). *Ocean, Environment, Climate Change and Human Migration*. International Organization for Migration. Available from: https://environmentalmigration.iom.int/sites/g/files/tmzbdl1411/files/ocean-environment-climate-change-migration_ScientificNotes_Oct2016_BD_ppp-12.pdf [Last accessed: 25th February 2025].

CMEMS (2024). *Area Averaged Sea Level: Global Ocean*. EU Copernicus Marine Service Information (CMEMS). Available from: https://marine.copernicus.eu/access-data/ocean-monitoring-indicators/time-series-mean-sea-level-anomalies-over-global-ocean-and-associated-trend [Last accessed: 5th February 2025].

EEA (2025). *Global and European Sea Level Rise*. European Environmental Agency. Available from: https://www.eea.europa.eu/en/analysis/indicators/global-and-european-sea-level-rise/past-trend-and-projected-change [Last accessed: 5th February 2025].

Freestone, D., & Çiçek, D. (2021). *Legal Dimensions of Sea Level Rise: Pacific Perspectives*. World Bank Group.

Glavovic, B. C., Dawson, R., Chow, W., Garschagen, M., Haasnoot, M., Singh, C., & Thomas, A. (2022). *Cross-Chapter Paper 2: Cities and Settlements by the Sea* (Climate Change 2022: Impacts, Adaptation and Vulnerability. Contribution of Working Group II to the Sixth Assessment Report of the Intergovernmental Panel on Climate Change, Issue). Cambridge University Press.

Gulev, S. K., Thorne, P. W., Ahn, J., Dentener, F. J., Domingues, C. M., Gerland, S., Gong, D., Kaufman, D. S., Nnamchi, H. C., Quaas, J., Rivera, J. A., Sathyendranath, S., Smith, S. L., Trewin, B., von Schuckmann, K., & Vose, R. S. (2021). *Changing State of the Climate System* (Climate Change 2021: The Physical Science Basis. Contribution of Working Group I to the Sixth Assessment Report of the Intergovernmental Panel on Climate Change, Issue). Cambridge University Press.

Hooijer, A., & Vernimmen, R. (2021). Global LiDAR Land Elevation Data Reveal Greatest Sea-Level Rise Vulnerability in the Tropics. *Nature Communications*, 12(1), 3592. https://doi.org/10.1038/s41467-021-23810-9.

Ietto, F., Conforti, M., Tolomei, C., & Cianflone, G. (2022). Village Relocation as Solution of the Landslide Risk, Is It Always the Right Choice? The Case Study of Cavallerizzo Ghost Village (Calabria, Southern Italy). *International Journal of Disaster Risk Reduction*, 81, 103267. https://doi.org/10.1016/j.ijdrr.2022.103267.

Magnan, A. K., Oppenheimer, M., Garschagen, M., Buchanan, M. K., Duvat, V. K. E., Forbes, D. L., Ford, J. D., Lambert, E., Petzold, J., Renaud, F. G., Sebesvari, Z., van de Wal, R. S. W., Hinkel, J., & Pörtner, H.-O. (2022). Sea Level Rise Risks and Societal Adaptation Benefits in Low-Lying Coastal Areas. *Scientific Reports*, 12(1), 10677. https://doi.org/10.1038/s41598-022-14303-w.

Marzeion, B., & Levermann, A. (2014). Loss of Cultural World Heritage and Currently Inhabited Places to Sea-Level Rise. *Environmental Research Letters*, 9(3), 034001. https://doi.org/10.1088/1748-9326/9/3/034001.

NASA (n.d.). *Sea Level*. Available from: https://climate.nasa.gov/vital-signs/sea-level/?intent=121 [Last accessed: 5th February 2025].

National Ocean Service (n.d). *How Much Oxygen comes from the Ocean?* National Oceanic and Atmospheric Administration. Available from: https://oceanservice.noaa.gov/facts/ocean-oxygen.html [Last accessed: 5th February 2025].

Oppenheimer, M., Glavovic, B. C., Hinkel, J., van de Wal, R., Magnan, A. K., Abd-Elgawad, A., Cai, R., Cifuentes-Jara, M., DeConto, R. M., Ghosh, T., Hay, J., Isla, F.,

Marzeion, B., Meyssignac, B., & Sebesvari, Z. (2019). *Sea Level Rise and Implications for Low-Lying Islands, Coasts and Communities* (IPCC Special Report on the Ocean and Cryosphere in a Changing Climate, Issue). Cambridge University Press.

Otosaka, I. N., Shepherd, A., Ivins, E. R., Schlegel, N. J., Amory, C., van den Broeke, M. R., Horwath, M., Joughin, I., King, M. D., Krinner, G., Nowicki, S., Payne, A. J., Rignot, E., Scambos, T., Simon, K. M., Smith, B. E., Sørensen, L. S., Velicogna, I., Whitehouse, P. L., & Wouters, B. (2023). Mass Balance of the Greenland and Antarctic Ice Sheets from 1992 to 2020. *Earth Syst. Sci. Data*, 15(4), 1597–1616. https://doi.org/10.5194/essd-15-1597-2023.

Reimann, L., Vafeidis, A. T., & Honsel, L. E. (2023). Population Development as a Driver of Coastal Risk: Current Trends and Future Pathways. *Cambridge Prisms: Coastal Futures*, 1, e14. https://doi.org/10.1017/cft.2023.3

UCCRN (2018). *The Future We Don't Want: How Climate Change Could Impact the World's Greatest Cities*. Urban Climate Change Research Network. Available from: https://www.c40.org/wp-content/uploads/2023/04/1789_Future_We_Dont_Want_Report_1.4_hi-res_120618.original-compressed.pdf

UN Secretary-General's Climate Action Team (2024). *Surging Seas in a Warming World: The Latest Science on Present-Day Impacts and Future Projections of Sea-Level Rise*. United Nations. Available from: https://www.un.org/sites/un2.un.org/files/slr_technical_brief_26_aug_2024.pdf

Vousdoukas, M. I., Athanasiou, P., Giardino, A., Mentaschi, L., Stocchino, A., Kopp, R. E., Menéndez, P., Beck, M. W., Ranasinghe, R., & Feyen, L. (2023). Small Island Developing States Under Threat by Rising Seas Even in a 1.5 °C Warming World. *Nature Sustainability*, 6(12), 1552–1564. https://doi.org/10.1038/s41893-023-01230-5.

WCRP Global Sea Level Budget Group (2018). Global Sea-Level Budget 1993–Present. *Earth System Science Data*, 10(3), 1551–1590. https://doi.org/10.5194/essd-10-1551-2018.

WEF (2019). *The Global Risks Report 2019*. World Economic Forum. Available from: https://www3.weforum.org/docs/WEF_Global_Risks_Report_2019.pdf [Last accessed: 5th February 2025].

6 Looking for clear skies

"I am not a climate refugee – I want to stress that, as I have been very privileged compared to people who lost everything in a disaster and truly had to start their lives over again". Zack Porter talks to us from his house in Vermont, United States, not too far from where he's originally from, Massachusetts.

Zack tells us the story of his childhood:

> As a teenager, I often struggled with mental health, and found myself most at peace in wild places. I knew from an early age that I wanted to dedicate my life to protecting more wild places for future generations. I just had to figure out how to do it best.

At the age of 18, Zack took his first step towards that goal by joining the United States Forest Service, where he worked in the Cascade Mountain Range in Washington – a region known for its towering peaks and lush forests. But as he spent more time working on public lands, he became increasingly aware that the federal government was failing to care for them as it should. "I wanted to be an advocate - to push for what needed to be improved", he says, "and the most natural next step was the nonprofit sector. This was the beginning of the long trail that led me to become the co-founder of a new organisation, Standing Trees".

Zack spent the following decade working in the Northern Rocky Mountains of Montana, advocating for stronger protections for forests, rivers, and wildlife.

> Montana drew me because of its vast wild landscapes. The city of Missoula, where I went to college, also gave me something deeper – a real sense of community. It's a place filled with people who share a passion for conservation and the outdoors, and I felt connected to something bigger than myself.

DOI: 10.4324/9781003493662-7

Montana grew to be the place where Zack truly felt at home. This was also the same place where he decided to build a family. However, in addition to the beauty and wildness of the place, the region is also prone to wildfires.

Particularly during the summers of 2016 and 2017, the wildfire impacts in the area grew immensely.

> My wife and I had just had our daughter, who was only a few months old, when we experienced months of dense smoke covering the entire city. Week after week, the smoke was so oppressive that we couldn't safely be outdoors, and it wasn't much safer inside. The sky alternated between gray and black, with deep red sunsets augmented by the smoke. Our throats burned, our eyes watered, and even with the windows shut tight and air purifiers running, there was no real escape.

What hit Zack was not only the physical discomfort, but also the realisation that this was becoming the new normal. He tells us

> While fire is a natural and necessary force for forest health across much of the western U.S., decades of suppression policies have disrupted that balance, creating conditions that make wildfires more intense and more frequent. Combined with the impacts of climate change – particularly the dwindling snowpack that once helped regulate moisture in the region – and the rapid expansion of communities into the "Wildland-Urban Interface", the result has been devastating for public health.

Further, Missoula, just like many cities in the western United States, is surrounded by mountains on all sides – inevitably, smoke collects in these areas from fires near and far, and gets trapped for a long time in the valleys, not blowing out until the weather pattern changes. Temperatures often climbed towards 100 degrees Fahrenheit, and during those dreadful summers, Zack and his family could not leave the house for a long time.

> The thought of abandoning our jobs and our community in Montana and starting over again somewhere new was daunting. But with a newborn, your perspective changes – and I could not handle the anxiety and dread of living with a similar level of smoke every summer.

Zack and his wife decided to move out of Missoula, and looked at Vermont as a possible alternative.

What Zack could not foresee is that, while they were moving away from one climate disaster, they would soon have to face another of a very different kind: floods. "In July of 2023, we experienced two days of heavy rain that cut off our city from surrounding areas and devastated many homes and businesses," he tells us. Precipitation has been intensifying in the state, and flooding has become a growing concern.

"I've thought a lot about our choice to move here since the flooding", Zack continues, acknowledging that the decision to settle in Vermont, a place they thought would offer safety from climate disasters, now feels ironic. Climate trends show that the Northeast, particularly Vermont, is experiencing greater warming than many other parts of the country, leading to more intense periods of rainfall and longer dry spells. As a result, flooding has become a major issue, and Vermont is now one of the states in the United States requiring the most assistance to recover from natural disasters.

> The irony is that we moved here thinking we could escape the worst, but soon learned we had moved to a place that's exceptionally flood-prone. On the one hand, we only have ourselves to blame for our decision. On the other hand, it also shows how there's probably no place anyone can go to escape the consequences of climate change.

Reflecting upon the underlying cause of it all, Zack highlights the crucial role of building communities in the right areas.

> What we often don't stop to think about is that we've located our modern communities in places that are exceptionally prone to disasters, with or without climate change. Take for instance the recent wildfires in Los Angeles [in January 2025]: LA was just a small city in 1900. Since then, it has expanded up into the mountains in areas where natural fires have been regularly burning for millennia. The same goes for flooding in my home of Montpelier, Vermont. The city I now live in is located in a floodplain at the confluence of two rivers. The changing climate, coupled with the draining of wetlands and degradation of forests over the past two centuries, has created a fragile situation for our community and many others like it across the eastern U.S. The question isn't if there will be another flood, but when.

At the moment, wildfires are not a major issue in the Northeast Unites States, and the United States Forest Service predicts that the likelihood of fire remains low in that region. Still, Zack's organisation, Standing Trees, works in coalition with groups across the country to shape federal policy on wildfire management. Science is clear that logging exacerbates wildfires, as the older a forest gets, the more resilient it becomes. The thick bark of bigger, older trees and the closed canopy of an established forest help keep the ground materials from drying out as quickly, making it less vulnerable to fires. But logging remains the dominant management method for forests across the United

States, with vast numbers of forests being cut down, often before they have the chance to reach their full ecological potential.

Zack believes that one of the most important, yet untold, stories of climate change, is related to how its effects are often deeply local. "We make decisions every day about where to allow houses or businesses to be built. These are decisions we should have been making more consciously," he says.

> Now, we are facing the reality of the choices we've made as a society. Climate change is partially to blame, but so are our decisions to build in vulnerable locations or with inappropriate techniques (such as failing to use fire-resistant materials, or building homes at ground level in a floodplain). Often it is the people with the least economic means who are hit the hardest by disasters. Instead, we need to build in safer areas, make our communities more resilient to fires and floods, and take steps to allow natural ecosystems to perform at their highest potential.

Zack stresses that we have natural resources that have been degraded to the point where they cannot perform the ecological functions they once did.

For those with the financial means, the effects of climate disasters may be less dire. In California, for example, the wealthiest people may have little trouble rebuilding after fires. However, those in lower socio-economic brackets face a much harder road to recovery. "The federal government barely provides enough help for people to recover after a disaster", Zack points out, "and, unfortunately, the situation is only worsening as disasters become more frequent and intense."

Zack recognises that not everyone has the choice to leave. In some places, he argues, there needs to be a halt on developing the areas that have been repeatedly devastated by fires or floods. He reflects on human nature: "After a trauma, people's attention spans and memories are short. They tend to forget what happened in the past", he says. "In Montpelier, Vermont, for instance, there was major flooding in 1927, but very few people alive today remember it".

In his organisation, Zack works to help people understand the environmental history of the land – what it used to be and what it could be in the future if we took a more sustainable approach to managing the landscape.

> Most people don't fully appreciate why our communities are at such great risk or pause to consider the wisdom of rebuilding the same way in the same places. Too often we decide to forget the trauma rather than investigate it. We tell ourselves that lightning never strikes the same place twice, he adds, yet this is obviously a myth. Flood prone and fire-prone areas will experience disasters again, and again, and again. We are ignoring the underlying causes of these calamities at our own peril.

Figure 6.1 Zack Porter, co-founder and executive director of Standing Trees, on a kayak.

Benefits and dangers of wildfires: Two sides of the same coin

When thinking of wildfires, one tends to associate them with uncontrollable, destructive forces. As a matter of fact, over the years, wildfires have generated severe damages and more recently, with their increasing frequency, have damaged ecosystems, destroyed infrastructure, and even generated casualties, a trend that is expected to worsen in the future (Huber, 2018). Yet, fires are an essential natural process which has long been shaping our landscapes and the evolution of biodiversity, contributing to forests' regeneration, benefitting conservation and soil fertility (Kelly et al., 2020; Merson, 2020). In a way, nature has never been extraneous to fires and has been finding its way to resilience in the face of these events.

Oddly, certain species – such as sequoias in the United States – strongly depend on fires as seed release and reproduction happens thanks to burnt soil, taking advantage of its high heat (Davare, 2022; WWF & BCG, 2020). Such a dichotomy sheds light on the debated nature of fires, showing two sides of the same coin: one demonstrating the potential ecological benefits, and one the ultimately dangerous damages.

On the one hand, controlled fires have long been used for land and ecosystem management, also in traditional practices from indigenous communities (Davare, 2022). However, at times, also suppressing fires in the ecosystems that require them may turn out to be a harmful strategy (Kelly et al., 2020). As a result, the integrity and rejuvenation properties of ecosystems may be reduced (DellaSala et al., 2022). Other human-made changes to landscapes, such as land clearing for industrial and urban development, have also been a contributing force to wildfires (Davare, 2022). All these factors have contributed to changing and shaping landscapes and creating somewhat unnatural conditions.

The unpredictable and uncontrollable nature of wildfires is now coupled with changing climate conditions and ever-expanding settlement into previously undeveloped areas – as a result of these factors, their impact is becoming increasingly unstable. Increasing temperatures and prolonged droughts are making it harder to distinguish between wildfires' benefits and their destructive force. While wildfires can occur naturally, such as from lightning strikes, they are mostly caused by human activity, with an estimate determining that 75% of recent wildfires are man-made (WWF & BCG, 2020). Regardless of what triggers them, climate change is making wildfires more frequent and intense by creating hotter and drier conditions, which naturally favour the occurrence of fires and fuel their spread (Jones et al., 2024). As a result, both nature and people are now at risk.

In recent years, wildfires and their devastating impacts have been dominating headlines worldwide, such as those recently struck in California and Australia. All around the world, from the Amazon to the Arctic, wildfires have been showing the dangerous side of the coin, demonstrating once again the obvious influence of climate change in these contexts. Each year, about 4 million square kilometres of land globally – that is approximately half the size of the United States – are affected by fires (European Space Agency, 2021). Between 1979 and 2013, the global fire season length increased on average by 19% (Jolly et al., 2015). As a result, the forest area burned by fires has increased considerably, with new data indicating a 5.4% annual increase over the period between 2001 and 2023 (MacCarthy et al., 2024). In this new picture, human safety remains a major consideration (Faison et al., 2023). In fact, since the beginning of the 20th century, an estimated 18 thousand people have been killed or injured by wildfires, even when relying on conservative data estimations (Jones et al., 2024). In addition, when wildfires strike, they tend to have direct consequences on the living conditions of surrounding communities. For instance, in 2017 alone, wildfires led to the evacuation of 550,000 people globally (Pronto et al., 2018). Further, property and infrastructure are impacted, and the damages are felt for years after these events. More than 2,200 structures were damaged or destroyed by the 2023 wildfire in Lahaina, Hawaii. Similarly, 2,600 homes were destroyed in the 2020 fire in the communities of Talent and Phoenix, Oregon (Calkin et al., 2023). These are just a few of the examples demonstrating the catastrophic effects of these fires. In addition, property values in areas affected by fire are found to decrease, in the case of the United States by 10–20% (Howard, 2014). According to the Centre for Research on the Epidemiology of Disasters, the wildfires disasters[1] that have been reported since 1911 – at least 495 – have been associated with nearly $200 billion in damages[2] (EM-DAT & CRED/UCLouvain, 2025). To put things into perspective, in the United States alone, in 2022, wildfires caused approximately $3.3 billion in damages (NOAA NCEI, 2025).

The adverse effects of wildfires are far-reaching, and go well beyond their flames (WWF & BCG, 2020; Huber, 2018). Air pollution from wildfire smoke, notably a mixture of hazardous air pollutants, has been somewhat underestimated and its health impacts understudied (Lohman et al., 2007). Indeed, black and orange dusty skies turning cities and human settlements into unliveable environments have become a common aspect of wildfires. It is estimated that 340 thousand premature deaths that are attributable to respiratory and cardiovascular issues occur due to wildfire smoke annually (WWF & BCG, 2020). Despite the global nature of the issue, socio-economic contexts play a role in the protection from such risks. As commonly the case, health impacts due to degraded air quality are disproportionately felt by lower-income communities with less access to public health services (Carmenta et al., 2021), while people in wealthier areas are found to access more information and are therefore able to take protective action during wildfire smoke events (Burke et al., 2025). Nevertheless, even wealthier areas, such as cities in North America, have been experiencing the intense effects of fire smoke, especially in the 2023 fire season (Burke et al., 2025).

Further to human impacts, direct contact with smoke and flames, and the destruction of habitats that follow fires, have been jeopardising the survival of wildlife (WWF & BCG, 2020). The modified fire regime is contributing to biodiversity loss, with more than 4,000 terrestrial and freshwater species found to be threatened by increased fire frequency and intensity (Kelly et al., 2020). Specific disastrous events have been catastrophic for local wildlife and biodiversity. For instance, 3 billion animals were killed or displaced during Australia's wildfire disasters of 2019–2020 (WWF, 2020). In Canada alone,

2023 wildfires burned almost 7.8 million hectares of tree cover (MacCarthy et al., 2024). Plant and animal species are also affected by fires as food chains are altered, resultingly disrupting ecosystems (Daly, 2019).

Beyond the ecological implications, wildfires are also directly contributing to climate change as they emit greenhouse gases into the atmosphere (Gajendiran et al., 2024). Despite not representing the largest burnt areas (10%) (FAO and UNEP, 2020), forests – in all biomes – are associated with an elevated share of CO_2 emissions from fires due to their high carbon storage capacity. It is estimated that almost a quarter of annual CO_2 emissions related to fires is released from these ecosystems, while a bigger share of fire-related CO_2 emissions (65%) is generated by savannahs (van der Werf et al., 2017).

Ultimately, wildfire risk has increased and continues to do so. The Intergovernmental Panel on Climate Change estimated that higher temperatures would contribute to increasing the frequency of wildfires and are expected to expand the annual fire ignition risk window by 10–30% (Field et al., 2007). This calls for a strong adaptation capacity so that we learn from nature itself in adapting to this scenario, while also coming to terms with what it means to live with wildfires. Building adaptive capacity remains a nuanced reflection of local contexts, where the realities, resources, and needs of a community determine their ability to adapt and build resilience (Paveglio et al., 2015). But when it comes to building resilience to fire events, whether natural or induced, adaptation becomes only part of the solution. Limiting exposure will also be essential to reduce the detrimental impacts of wildfires (Huber, 2018). Yet, the history of wildfires and the impacts that are being felt are showing that communities have been populating areas that have been traditionally fire prone, but that now are even more so (Calkin et al., 2023). Although there is no single solution to this challenge, one thing is certain: addressing the root causes of wildfires is essential to reducing fire activity. And this can only be achieved by cutting greenhouse gas emissions and disrupting the fire-climate feedback loop (MacCarthy et al., 2024).

Notes

1 Incidents that killed at least 10 people or affected more than 100.
2 Amount of damages adjusted for inflation – see EM-DAT Documentation on Economic Adjustment.

Bibliography

Burke, M., Patel, L., Yu, J., & Chinthrajah, S. (2025). *What We Know About the Health Effects of Wildfire* [Interview]. StanfordReport. Available from: https://news.stanford.edu/stories/2025/01/assessing-wildfire-health-risks [Last accessed: 27th February 2025].

Calkin, D. E., Barrett, K., Cohen, J. D., Finney, M. A., Pyne, S. J., & Quarles, S. L. (2023). Wildland-Urban Fire Disasters Aren't Actually a Wildfire Problem.

Proceedings of the National Academy of Sciences, 120(51), e2315797120. https://doi.org/10.1073/pnas.2315797120.

Carmenta, R., Cammelli, F., Dressler, W., Verbicaro, C., & Zaehringer, J. G. (2021). Between a Rock and a Hard Place: The Burdens of Uncontrolled Fire for Smallholders Across the Tropics. *World Development*, 145, 105521. https://doi.org/10.1016/j.worlddev.2021.105521.

Daly, N. (2019). *What the Amazon Fires Mean for Wild Animals*. National Geographic. Available from: https://www.nationalgeographic.com/animals/article/how-the-amazon-rainforest-wildfires-will-affect-wild-animals# [Last accessed: 27th February 2025].

Davare, R. (2022). *The Impact of Wildfires on Biodiversity and the Environment*. Earth.Org. Available from: https://earth.org/impact-of-wildfires/ [Last accessed: 27th February 2025].

DellaSala, D. A., Baker, B. C., Hanson, C. T., Ruediger, L., & Baker, W. (2022). Have western USA Fire Suppression and Megafire Active Management Approaches Become a Contemporary Sisyphus? *Biological Conservation*, 268, 109499. https://doi.org/10.1016/j.biocon.2022.109499.

EM-DAT & CRED/UCLouvain. (2025). *Public EM-DAT*. Available from: www.emdat.be [Last accessed: 27th February 2025].

European Space Agency. (2021). *Multi-Decade Global Fire Dataset Set to Support Trend Analysis*. Available from: https://climate.esa.int/de/news-events/multi-decade-global-fire-dataset-set-support-trend-analysis/ [Last accessed: 27th February 2025].

Faison, E. K., Masino, S. A., & Moomaw, W. R. (2023). The Importance of Natural Forest Stewardship in Adaptation Planning in the United States. *Conservation Science and Practice*, 5(6), e12935. https://doi.org/10.1111/csp2.12935.

FAO and UNEP, (2020). *The State of the World's Forests 2020: Forests, Biodiversity and People*. Available from: https://doi.org/10.4060/ca8642en [Last accessed: 12 March 2025].

Field, C. B., Mortsch, L. D., Brklacich, M., Forbes, D. L., Kovacs, P., Patz, J. A., Running, S. W., & Scott, M. J., (2007). North America. In: M. L. Parry, O. F. Canziani, J. P. Palutikof, P. J. van der Linden, & C. E. Hanson (Eds.) *Climate Change 2007: Impacts, Adaptation and Vulnerability. Contribution of Working Group II to the Fourth Assessment Report of the Intergovernmental Panel on Climate Change*. Cambridge University Press, pp. 617–652.

Gajendiran, K., Kandasamy, S., & Narayanan, M. (2024). Influences of Wildfire on the Forest Ecosystem and Climate Change: A Comprehensive Study. *Environmental Research*, 240, 117537. https://doi.org/10.1016/j.envres.2023.117537.

Howard, P. (2014). *Flammable Planet: Wildfires and the Social Cost of Carbon*. Environmental Defense Fund, Institute for Policy Integrity, Natural Resource Defense Council. Available from: https://costofcarbon.org/files/Flammable_Planet__Wildfires_and_Social_Cost_of_Carbon.pdf [Last accessed: 27th February 2025].

Huber, K. (2018). *Resilience Strategies for Wildfire*. Center for Climate and Energy Solutions. Available from: https://www.c2es.org/wp-content/uploads/2018/11/resilience-strategies-for-wildfire.pdf [Last accessed 12th March 2025].

Jolly, W. M., Cochrane, M. A., Freeborn, P. H., Holden, Z. A., Brown, T. J., Williamson, G. J., & Bowman, D. M. J. S. (2015). Climate-Induced Variations in Global Wildfire Danger from 1979 to 2013. *Nature Communications*, 6(1), 7537. https://doi.org/10.1038/ncomms8537.

Jones, M. W., Kelley, D. I., Burton, C. A., Di Giuseppe, F., Barbosa, M. L. F., Brambleby, E., Hartley, A. J., Lombardi, A., Mataveli, G., McNorton, J. R., Spuler, F. R., Wessel, J. B., Abatzoglou, J. T., Anderson, L. O., Andela, N., Archibald, S., Armenteras, D., Burke, E., Carmenta, R., & Xanthopoulos, G. (2024). State of Wildfires 2023–2024. *Earth System Science Data*, 16(8), 3601–3685. https://doi.org/10.5194/essd-16-3601-2024.

Kelly, L. T., Giljohann, K. M., Duane, A., Aquilué, N., Archibald, S., Batllori, E., Bennett, A. F., Buckland, S. T., Canelles, Q., Clarke, M. F., Fortin, M.-J., Hermoso, V., Herrando, S., Keane, R. E., Lake, F. K., McCarthy, M. A., Morán-Ordóñez, A., Parr, C. L., Pausas, J. G., & Brotons, L. (2020). Fire and Biodiversity in the Anthropocene. *Science*, 370(6519), eabb0355. https://doi.org/10.1126/science.abb0355.

Lohman, D. J., Bickford, D., & Sodhi, N. S. (2007). The Burning Issue. *Science*, 316(5823), 376–376. https://doi.org/10.1126/science.1140278.

MacCarthy, J., Richter, J., Tyukavina, S., Weisse, M., & Harris, N. (2024). *The Latest Data Confirms: Forest Fires Are Getting Worse*. World Resource Institute. Available from: https://www.wri.org/insights/global-trends-forest-fires [Last accessed: 27th February 2025].

Merson, J. (2020). Climate Change and Wildfires: Lessons from Australia's Blue Mountains. *Crossroads*. Available from: https://iucn.org/crossroads-blog/202012/climate-change-and-wildfires-lessons-australias-blue-mountains [Last accessed: 27th February 2025].

NOAA NCEI (2025). *U.S. Billion-Dollar Weather and Climate Disasters*. https://doi.org/10.25921/stkw-7w73.

Paveglio, T., Moseley, C., Carroll, M., Williams, D., Davis, E., & Fischer, A. (2015). Categorizing the Social Context of the Wildland Urban Interface: Adaptive Capacity for Wildfire and Community "Archetypes". *Forest Science*, 61, 298–310. https://doi.org/10.5849/forsci.14-036.

Pronto, L. N., Byambasuren, O., & Goldammer, J. G. (2018). *Global Wildland Fire Network Bulletin* (Annual Global Wildland Fire Fatalities and Damages Report, Issue 23). GFMC, IWPM, UNISDR. Available from: https://gfmc.online/wp-content/uploads/GFMC-Bulletin-01-2018.pdf [Last accessed: 27th February 2025].

van der Werf, G. R., Randerson, J. T., Giglio, L., van Leeuwen, T. T., Chen, Y., Rogers, B. M., Mu, M., van Marle, M. J. E., Morton, D. C., Collatz, G. J., Yokelson, R. J., & Kasibhatla, P. S. (2017). Global Fire Emissions Estimates During 1997–2016. *Earth System Science Data*, 9(2), 697–720. https://doi.org/10.5194/essd-9-697-2017.

WWF (2020). *Australia's 2019-2020 Bushfires: The Wildlife Toll*. Available from: https://assets.wwf.org.au/image/upload/Animals_Impacted_Interim_Report_24072020_final?_a=ATO2Bcc0 [Last accessed: 27th February 2025].

WWF & BCG. (2020). *Fires, Forests, and the Future: A Crisis Raging Out of Control?* Available from: https://wwfeu.awsassets.panda.org/downloads/wwf_fires_forests_and_the_future_report.pdf [Last accessed: 27th February 2025].

7 Lost memories from a flood

"Any extreme weather event is lived very differently, depending on the people you talk to - and this is often a question of income". Vincy Abraham is a young Indian woman, a climate activist, and a community worker. She grew up in the slums of Mumbai, where she directly experienced what she calls a life without resources.

> Basic things like access to water and personal latrines were never to take for granted, and summers would get unbearably hot, as all roofs were made of asbestos. This is why, whenever something like a flood happens, you can easily imagine how shacks are differently impacted than stable buildings.

When talking about her career as a policy expert on the impact of climate change on rural communities, her eyes lighten up, and she explains how it's so key to support communities, such as the ones she was born in, to adapt to the consequences of climate change. "You see, very often we forget how climate change impacts first and foremost people living in poverty - and when designing large-scale policies, we should keep in mind questions of climate justice".

Vincy was able to witness this through her teenage eyes, when her family was heavily impacted by the devastating Mumbai flood of July 2005. In one day, the megacity received the average amount of rainfall for the entire season which, combined with high-tides, set off an unprecedented flood.

"I still remember every detail of that day - it was 26 July - just as if it happened a few days ago" she says. "I was around 15 at the time – a very important age in India as this is when all teenagers prepare for their tenth grade exam, allowing us to access college and university".

At the time, she said, she felt relatively lucky because her family had just moved to better housing, going from a typical slum-house to a ground-floor apartment on a two-floor building. However, she also knew she was one of

DOI: 10.4324/9781003493662-8

the few students in her school to come from a slum background – something which, at the time, made her relatively ashamed. "I am now proud of my background and belonging, as it shaped who I am right now" she says. "But it wasn't always like this. As a teenager, I felt different - I was one of the few children from the slums who was studying, and this always made me feel out of place", Vincy adds.

The day of the flood started just like any other normal day – she woke up in the morning, got ready, and went to school. On the way there, she noticed how rain was pouring in a weird way. "It was much stronger than normal, nothing I had ever seen before" she tells us.

After a few hours of class, the rain kept pouring, at a rate never witnessed before by most people. For this reason, teachers and managers decided to close the school for security purposes, as a flood was expected. "At that time, the first feeling I had was that I really wanted to go home and be close to my family," she says, a shadow passing through her eyes.

After a failed attempt to get some buses to reach the school, the teachers decided to let the oldest pupils leave if they wanted to. She didn't think twice – and this was when her odyssey began.

Vincy's school was on the west side of Mumbai, while her house was on the east side, so she knew she was in for a long walk. As soon as she left the school, she saw a couple in a car that was getting completely submerged by water. A group of older students was trying to help the couple open the doors of the car, while some people were standing close to the drainage so the car would not fall in that direction.

Seeing she couldn't do much more of what others were already doing, she kept going, slowly getting to the highway. There, she had a choice: take a safer street where people needed to swim or pass through a street completely packed with abandoned vehicles, which were incessantly moved by the passing of water. "It was then when I remember that nobody had ever taught me how to swim", she says. There was no other choice: she needed to walk through the vehicles.

All of the sudden, her foot got stuck into something hidden beneath the muddy water. She started panicking and looking around, seeing a lady who was walking in the opposite direction. As soon as the lady saw her, she immediately walked on her side of the street, calming her down and helping her to free her foot. This happened very suddenly, but until today she doesn't know what she would have done without the help of this lady. "It was a long trip; I think two or three hours," she says. "But most of all, what I remember, besides the fear, is the solidarity of the people around me".

It was only once she arrived home that she realised how her life was meant to change forever. When her mum opened the door, she saw how the whole house was already flooded with water. The first concern was to remove water from all the appliances and unplug the sockets. While she was helping her mum doing so, she noticed how her mum all of the sudden started grabbing a few of the best clothes they had, and putting them on the bed.

> I was about to ask my mum what happened, but a loud noise distracted me. Turning around, I saw our washing machine freely moving across the entry hall, brought by a very strong current, and reaching the front room. It was then that my mum said we had no other choice but to move upstairs.

And this is what they did - they moved upstairs and were stuck there for a few days, among neighbours and other family friends who sought refuge. It was only after three days of rain that they were able to reconnect with the rest of the family – her sister and her dad who were in very different parts of the city when the flood happened. "I remember we moved to our uncle's house, which at least felt a little bit more like home. But when we finally could return to our old apartment, everything was destroyed".

Despite trying to clean everything and throwing away furniture and appliances, the house walls were still very wet, and they had to sleep on the floor for weeks. But what most concerned her was returning back to school: she soon realised that all the other kids had not been as impacted as her, as they came from better households, which were either on higher floors or had some form of protection against flooding. She explains

> I had a lot of work to do, with incoming exams, and yet my parents were talking about having to move apartments but having no money for it, we had no furniture, and the house was just not healthy to live in.

It was around that time that she started noticing very strong joint pain, and found out she caught a bacterial infection called leptospirosis, which probably came from the cuts on her leg in contact with the mouldy water.

"I didn't go to school for many months afterwards", she tells, "and I became very afraid of not being able to take and pass the exams. Somehow, I still managed. But my life and my family's life was never as before".

Her parents, who were still not doing well financially, were hit by a very hard toll. It took a while for them to be able to afford another apartment and get out of the rental contract of the first one. And the way they lived through this was very different from most of her classmates, who were not able to understand what a flood could do to poorer, less fortunate people, and how it could impact them.

Climate change, Vincy says, left both tangible and intangible effects on her. What is tangible is that she lost most of her clothes, photos, and objects of when she was young. What's intangible is the illness she caught, as well as all these hurtful memories from her youth that she could never really let go of. As she was getting married, a few months ago, she mentioned how her husband asked her to look for some photos of when she was young, to play a small movie during the wedding, but sadly she had none, she continues.

> This was the first, but unfortunately not the last intense flood I remember. I think floods of this magnitude started more or less 15 years ago, and now they are more and more frequent, and every time I live through one, I think about how many families are affected by it.

Figure 7.1 Vincy Abraham, policy expert, climate activist, and community worker, on a ship deck.

When asked about whether she thinks the city improved and learned, she says,

> You know, through time, I think we, as citizens of Mumbai, understood collectively that climate change is here to stay. By bonding together, and helping each other, we became much more resilient. But while we grew stronger, the city's infrastructure didn't, and is not as resilient as it would need to be to protect everyone – and not only the most privileged.

Heavy rainfalls and floods threatening communities

From a resources perspective, one might say that there is no such thing as too much water. Yet, in a changing climate, even rain, often considered a blessing, may turn into a curse. As a matter of fact, with rain come both positive and negative consequences. While the positives are broadly known, the detrimental effects of rainfalls are often overlooked and hard to prevent. As emphasised by recent headlines, both urban and rural areas around the world have been increasingly battling with this challenge. However, this is pointing to an issue that goes well beyond the mere amount of rainfall, all the way to matters of water management, landscape planning, and social policy.

With climate change, rising temperatures tend to intensify the earth's water cycle. Thus, evaporation increases as well, leading to more water in the air (Trenberth, 2011). Every decade, as the earth warms, atmospheric water vapour, which supplies water for precipitation, increases by 1–2% (Douville et al., 2021). As a result, we can expect more frequent and intense rainfalls and storms on the one hand, while also intense drought events on the other (Trenberth, 2011).

An increase in frequency and intensity of heavy rainfalls has already been witnessed by many regions of the world, with the IPCC reporting a likely increase occurred globally over the majority of land regions (Seneviratne et al., 2021). In addition, since 1990, nine out of ten of the most catastrophic disasters have been related to water (UNISDR, 2014). In particular, almost half of these disasters were flood events, making flooding the most common disaster (Doocy et al., 2013; Institute for Economics & Peace, 2020).

Floods, which happen when water cannot drain away properly, are the most detrimental consequence of heavy rainfalls. They may also result from high tides, and in some areas from sea level rise and eruptions of glacial lakes. The way landscapes have been developed strongly influences the outcomes of flood events. Environmental features such as soil type, the presence of vegetation, the use of concrete and integration of drainage systems determine the ability of the surface to absorb water, with dryer and harder surfaces often failing to absorb enough (Doocy et al., 2013; Environmental Defense Fund, 2024). Ecosystem degradation, known to worsen the condition of the soil, is in fact one of the main causes of increasing challenges in relation to water resources management (WWAP, 2018). In addition, climatic conditions play an important role. For instance, if a heatwave occurs before heavy precipitation, the soil will be so dry that the water will be hardly absorbed, therefore increasing the risk of flooding (Environmental Defense Fund, 2024).

In particular, urban areas, often characterised by the presence of pavements, non-naturalised areas, and inadequate stormwater management, are becoming increasingly prone to floods (Rubinyiross & Eisenberg, 2023). As urbanisation continues to rise, so does the density of infrastructures and their vulnerability to the consequences of heavy rainfalls (Donnor & Rodríguez, 2011). For instance, coastal and dense cities, such as Mumbai, experience frequent flood events, resulting from heavy rainfall and a combination of surface and tidal flooding, in addition to channel overtopping (Eldho et al., 2018).

In addition to being the most common natural disasters, floods are also those leading to most casualties (Doocy et al., 2013). People's livelihoods are the first to be impacted as extreme events of this kind are found to impact both economic and social stability (Kotz et al., 2023). It is estimated that almost one quarter of the world population are directly exposed to a 1-in-100-year flood event[1] (Rubinyiross & Eisenberg, 2023). The harmful impacts of floods include mortality, loss or damage of infrastructure and property leading to displacement and migration, damage to crops therefore disrupting livelihoods, as well as trauma and psychosocial effects. In the longer term, infectious diseases from contact with waters are found to lead to health issues (APFM, 2013; Doocy et al., 2013). Globally, Asia is the region most affected by floods, where about half of all flood-related fatalities have happened in the last quarter of the 20th century (Doocy et al., 2013).

As with many impacts of climate change, vulnerable communities are the ones that result most affected by floods. Data indicates that out of ten people exposed to high flood risks, four live in poverty, amounting to 780 million people living on less than $5.50 per day (Salhab & Rentschler, 2020). Vulnerable communities and people living in poverty, while often living in flood-prone areas, may not have access to flood-protection infrastructure and to the resources needed to evacuate or shelter themselves, which generally are more accessible in high-income neighbourhoods (Ermagun et al., 2024; Lindersson et al., 2023).

Flooding has also been associated with substantial economic losses. The year 2021 witnessed some of the most extreme flooding from Europe to Asia, with related news dominating headlines globally. In the same year, flooding resulted as the most costly extreme event amounting to $105 billion (AON, 2023). Floods in Pakistan caused more than $15 billion in damages, killing more than 1,700 people and displacing around 8 million people (The Government of Pakistan et al., 2022). The flash floods that occurred in western Germany and bordering countries in 2021 are considered the costliest flood event in history, leading to losses of $45 billion (Munich, n.d.).

In India, heavy rainfalls have been happening for many years, almost becoming commonplace. But floods have also become increasingly common, the unpredictability of the monsoon has increased and so has the rain it brings. Looking at the history of floods in the country, the most disastrous ones have occurred after the year 2000 (Sahu et al., 2023). Since 2012, heavy rainfalls have increased by 85% in the country. As a result, the impacts of these events have also intensified (World Bank, 2023).

According to climate models, the picture at the horizon is not promising. Particularly, in dry regions, precipitation is following a "it never rains, but it pours" pattern (Tabari, 2020), meaning that while overall rainfall might decrease, when it does rain, it will be very intense, even if not necessarily more frequent. Conversely, in wetter areas, more available moisture can lead to even more intense rain and floods, increasing in frequency.

But in the face of increasing extreme events such as heavy rainfalls, it is important to ensure that they do not culminate into actual disasters. If mitigating the impacts of floods requires implementing policies to the support of all those affected, and especially vulnerable people, preventing and managing floods relies on radical changes in the ways landscapes are developed. Once again, nature offers part of the solution, and an opportunity to complement purely conventional engineering tools through which water-related challenges have been traditionally addressed. Concepts such as "sponge cities",[2] referring to urban planning models that use abundant natural landscapes to absorb, store, and purify rainwater (Harrisberg, 2022; Nguyen et al., 2019), are becoming more and more of an inspiration, providing a glimpse of hope for the future.

Notes

1 A flood that statistically has a 1% chance of occurring in any given year.
2 Concept developed in 2014 by the People's Republic of China to address urban surface-water flooding and other water-related challenges by making use of blue and green spaces in 30 pilot cities (Ka Shu Chan et al., 2018).

Bibliography

AON (2023). *Weather, Climate and Catastrophe Insight*. Available from: https://www.aon.com/getmedia/f34ec133-3175-406c-9e0b-25cea768c5cf/20230125-weather-climate-catastrophe-insight.pdf [Last accessed: 30th January 2025].

APFM (2013). *What Are the Negative Social Impacts of Flooding?* Associated Programme on Flood Management. Available from: https://www.floodmanagement.info/what-are-the-negative-social-impacts-of-flooding/ [Last accessed: 30th January 2025].

Donnor, W., & Rodríguez, H. (2011). *Disaster Risk and Vulnerability: The Role and Impact of Population and Society*. Population Reference Bureau. Available from: https://www.prb.org/resources/disaster-risk/[Last accessed: 30th January 2025].

Doocy, S., Daniels, A., Murray, S., & Kirsch, T. D. (2013). The Human Impact of Floods: a Historical Review of Events 1980-2009 and Systematic Literature Review. *PLoS Curr,* 5. https://pubmed.ncbi.nlm.nih.gov/23857425/

Douville, H., Raghavan, K., Renwick, J., Allan, R. P., Arias, P. A., Barlow, M., Cerezo-Mota, R., Cherchi, A., Gan, T. Y., Gergis, J., Jiang, D., Khan, A., Pokam Mba, W., Rosenfeld, D., Tierney, J., & Zolina, O. (2021). *Water Cycle Changes* (Climate Change 2021: The Physical Science Basis. Contribution of Working Group I to the Sixth Assessment Report of the Intergovernmental Panel on Climate Change Issue). Cambridge University Press.

Eldho, T. I., Zope, P. E., & Kulkarni, A. T. (2018). Chapter 12 - Urban Flood Management in Coastal Regions Using Numerical Simulation and Geographic Information System. In: P. Samui, D. Kim, & C. Ghosh (Eds.) *Integrating Disaster Science and Management*. Elsevier, pp. 205–219. https://doi.org/10.1016/B978-0-12-812056-9.00012-9

Environmental Defense Fund. (2024). *Why Are Floods Hitting More Places and People?* Available from: https://www.edf.org/why-are-floods-hitting-more-places-and-people [Last accessed: 30th January 2025].

Ermagun, A., Smith, V., & Janatabadi, F. (2024). High Urban Flood Risk and No Shelter Access Disproportionally Impacts Vulnerable Communities in the USA. *Communications Earth & Environment*, 5(1), 2. https://doi.org/10.1038/s43247-023-01165-x.

Harrisberg, K. (2022). *What Are 'Sponge Cities' and How Can They Prevent Floods?* World Economic Forum. Available from: https://www.weforum.org/stories/2022/03/sponge-cities-nature-tackle-climate-floods-urban/ [Last accessed: 30th January 2025].

Institute for Economics & Peace. (2020). *Ecological Threat Register 2020: Understanding Ecological Threats, Resilience and Peace*. Available from: http://visionofhumanity.org/reports [Last accessed: 30th January 2025].

Kotz, M., Lange, S., Wenz, L., & Levermann, A. (2023). Constraining the Pattern and Magnitude of Projected Extreme Precipitation Change in a Multimodel Ensemble. *American Meteorological Society*, 97–111. https://doi.org/10.1175/JCLI-D-23-0492.1.

Lindersson, S., Raffetti, E., Rusca, M., Brandimarte, L., Mård, J., & Di Baldassarre, G. (2023). The Wider the Gap between Rich and Poor the Higher the Flood Mortality. *Nature Sustainability*, 6(8), 995–1005. https://doi.org/10.1038/s41893-023-01107-7.

Munich, R. E. (n.d.). *Flood Risks on the Rise: Greater Loss Prevention Is Needed*. Available from: https://www.munichre.com/en/risks/natural-disasters/floods.html [Last accessed: 30th January 2025].

Nguyen, T. T., Ngo, H. H., Guo, W., Wang, X. C., Ren, N., Li, G., Ding, J., & Liang, H. (2019). Implementation of a Specific Urban Water Management - Sponge City. *Science of the Total Environment*, 652, 147–162. https://doi.org/10.1016/j.scitotenv.2018.10.168.

Rubinyiross, S., & Eisenberg, R. (2023). *Decoding Urban Flood Risk and Democratizing Risk Assessment*. World Bank. Available from: https://blogs.worldbank.org/en/sustainablecities/decoding-urban-flood-risk-and-democratizing-risk-assessment#:~:text=Urban%20areas%20are%20increasingly%20prone,risk%20to%20lives%20and%20livelihoods [Last accessed: 30th January 2025].

Sahu, S., Sahu, B., & Ramteke, S. (2023). Floods Disaster in India, Mitigation and Their Impacts. *Sustainability Agri Food and Environmental Research*, 12(2024). Available from: https://www.researchgate.net/publication/371069486_Floods_disaster_in_India_mitigation_and_their_impactshttps://www.researchgate.net/publication/371069486_Floods_disaster_in_India_mitigation_and_their_impacts [Last accessed: 30th January 2025].

Salhab, M., & Rentschler, J. (2020). *People in Harm's Way: Flood Exposure and Poverty in 189 Countries*. World Bank. Available from: http://hdl.handle.net/10986/34655 [Last accessed: 30th January 2025].

Seneviratne, S. I., Zhang, X., Adnan, M., Badi, W., Dereczynski, C., Di Luca, A., Ghosh, S., Iskandar, I., Kossin, J., Lewis, S., Otto, F., Pinto, I., Satoh, M., Vicente-Serrano, S. M., Wehner, M., & Zhou, B. (2021). *Weather and Climate Extreme Events in a Changing Climate* (Climate Change 2021: The Physical Science Basis. Contribution of Working Group I to the Sixth Assessment Report of the Intergovernmental Panel on Climate Change, Issue). Cambridge University Press.

Tabari, H. (2020). Climate Change Impact on Flood and Extreme Precipitation Increases With Water Availability. *Scientific Reports*, 10(1), 13768. https://doi.org/10.1038/s41598-020-70816-2.

The Government of Pakistan, Asian Development Bank, European Union, United Nations Development Programme, & Bank, W. (2022). *Pakistan Floods 2022: Post-Disaster Needs Assessment*. Available from: https://www.preventionweb.net/media/84983/download?startDownload=20250130 [Last accessed: 30th January 2025].

Trenberth, K. (2011). Changes in Precipitation With Climate Change. *Climate Research*, 47, 123–138. https://doi.org/10.3354/cr00953.

UNISDR (2014). *Annual Report*. United Nations Office for Disaster Risk Reduction. Available from: https://www.preventionweb.net/files/42667_unisdrannualreport2014.pdf [Last accessed: 30th January 2025].

World Bank. (2023). *India: Managing the Complex Problem of Floods and Droughts*. Available from: https://www.worldbank.org/en/news/feature/2023/08/17/india-managing-the-complex-problem-of-floods-and-droughts [Last accessed: 30th January 2025].

WWAP (2018). *The United Nations World Water Development Report 2018: Nature-Based Solutions for Water*. United Nations World Water Assessment Programme/UN-Water - UNESCO. Available from: https://unesdoc.unesco.org/ark:/48223/pf0000261424 [Last accessed: 30th January 2025].

8 The impermanence of "Eternal ice"

"My country's landscape and culture are slowly changing". The vivid eyes of Bekbolat turn to a point on the horizon, changing into a frowned, concerned expression.

Bekbolat was born in the city of Olgij, a far-off land of high mountains (the Mongol Altai, with peaks higher than four thousands metres) as well as torrents and glaciers, in the most western part of the country of Mongolia, inhabited by a Kazakh minority who herd yaks and goats and still hunt with trained eagles. He has Kazakh roots and a long, nomadic tradition spanning through centuries. His city was traditionally a gateway to the region's diverse landscape, including large glaciers, and vast steppes.

Today, Bekbolat runs a small tourist company, Altai expeditions, where he takes travellers from all around the world up on the Mongolian mountains. Every day, he meets all kinds of hikers – young groups of friends training for their expeditions, older couples in love with the incredible landscape, and scientists interested in Mongolia's glaciers.

> There is a word – translated it would mean eternal ice – that we use to call our glaciers. But during my expeditions I have seen with my own eyes how these glaciers are not eternal anymore – if anything, they are halfening in front of our eyes. And this is of course just the beginning.

Bekbolat has always been interested in learning more about the delicate equilibrium of the natural ecosystem. This also led him to move to the capital, Ulaanbaatar, to study ecology and become a primary school teacher. In 2009, he decided that his life needed a twist – he wanted to combine his passion for nature with a job that allowed him to learn about the world – and started to learn English with the objective to enter the world of tourism. He tells us:

> It was shortly after I had started organising my first tourist expeditions that I was called by a group of Russian, Mongolian and Japanese scientists, interested in measuring the alterations in size of the glacier Potanin, which is the largest in Mongolia.

DOI: 10.4324/9781003493662-9

At first, he was impressed by how people from all around the world were interested in a phenomenon he thought was a local problem. He was aware of how many of Mongolia's glaciers had started melting at a much faster rate than normal, and this was already impacting the landscape as well as water availability in the region.

One expedition gave rise to many others. While on these expeditions with scientists, he learned how the Potanin glacier was getting smaller every year, and at a faster rate. "What I understood then was that the melting rate was exponential, not something progressive. The Russian scientist informed me that, when he first came, the Potanin glacier was 20 km long – now there are only 10km left".

One of the most surprising things happened right in one of his expeditions. One day, he stumbled upon something dark, brown, and sharp which was hiding underneath a thin layer of melted ice on the side of the glacier. He picked it up, curious, and noticed how it looked like an arrow. After cleaning it with his fingers, he noted how it was very carefully crafted and probably made of animal bones. "This was when I remember my parents telling me ancient stories about our culture, and how once upon a time our ancestors used to hunt animals in the region of the glaciers".

"I decided to bring the small arrow back to the local museum, and they were able to trace it back to more than a thousand years ago, to the Genghis Khan's era", he smiles. "At first, I was extremely happy about such a discovery. How amazing it was to say that I could find and hold an arrow which belonged to a former ancestor, already a craftsman and probably a skilled hunter!". However, after a while, Bekbolat started to realise what this meant. This was a tangible proof that the glacier was now melting at an unprecedented rate, up to the point of unveiling rare artefacts from his ancestors. Bekbolat states:

> The glacier had been keeping these precious tools from our ancestors intact up until this time. Now, with its melting, artefacts are slowly getting destroyed by the mass of earth and debris. This is also further evidence that there's no going back from the melting, as what is lost will not be recovered.

There is a high level of awareness among Bekbolat's people about the dangers posed by climate change – he assures us –, for instance, among his brothers, his family, and friends, and also among the older generations. "People", he tells us, "already feel climate change on their skin: they say the weather is not stable and that there are very dry summers. They just don't know what to do about it".

Bekbolat's two brothers live in the countryside and have a lot of animals, among which are yaks and cows. "Last summer was quite dry and there wasn't a lot of rain, so the grass did not grow", he says. "Nomadic people in this area have started importing grass from other regions – areas either in the north or the east of Mongolia where there is still more rain. This is key if they want to keep their cattle alive".

But Bekbolat's biggest concern is abound the availability of water: Mongolia is a nomadic country and people rely on animals to make ends meet, but the scarcity of water is threatening the culture of the landscape.

> The Mongolian Altai region has a big river named Holt [or/Khovd/], and all the small rivers feed into this big one. Normally, the regular melting of a part of the glaciers in the summer feeds into these rivers. However, if the glaciers' reserve continues to diminish at this rate, one day there won't be enough water to fill our rivers – and the first ones to realise this will be the animals.

> "Already today, nomadic life is slowly disappearing" he sighs.

> People are moving to the city as there are less and less animals available and feeding them costs too much. Young people are starting to realise that they would need to learn something different. Imagine what this would mean with rivers slowly drying up.

Bekbolat is also aware of how the unique heritage and culture of nomads and the beautiful nature of the steppe are two factors that are attracting tourists to Mongolia, a country that opened up to the rest of the world only around 30 years ago with the end of the USSR. Preserving Mongolia's natural landscape, therefore, is truly critical for the country. He says:

> With my work, I try to communicate to the tourists about the natural wonders that Mongolia holds – our customs, our traditions, and our landscape. Aside from this, I also run a few volunteering projects. For instance, in 2015 I founded an NGO called "Achievable Solutions", with the aim to improve local communities, for instance by providing drinking water to families affected by floods, or by focusing on educational projects for young children, including environmental education.

Education is key to Bekbolat, who is committed to preserve Mongolia's precious natural and cultural heritage, and to help people around him. He believes that young Mongolians should progressively be made aware of the threats posed by climate change to better understand how to tackle it.

> While it was incredible to find the small arrow belonging to my ancestors on the glacier, this made me realise even more about the challenges our future will hold. With our glaciers melting, we might be rediscovering our past, but we are also losing sight of our future.

Figure 8.1 Bekbolat Bugibay, the owner of Altai expeditions, wearing traditional Kazakh clothing.

Bekbolat is now looking at an eagle, flying behind the summit of a mountain. "And I want to do anything possible to preserve both".

The ripple effect of melting glaciers

Today, glaciers, ice caps, and sheets cover around 10% of land worldwide (IPCC, 2019; NSIDC, 2024), storing 70% of the Earth' freshwater (USGS, 2019). High in the mountains, above the clouds, large glacier systems dominate some of the worlds' highest peaks and valleys. There, the ice melts and runs down the slopes from the glaciers' tongue, while being replenished from the ice flow in the upper part. However, these natural systems are being increasingly disrupted due to anthropogenic climate change, with about half of all mountain glaciers expected to melt by the end of the century (Harvey, 2023). One question therefore arises naturally: what will happen if all glaciers melt away?

Glaciers are much more than the accumulations of snow and ice they originate from: they are incredibly important sources of freshwater supply. As such, they feed rivers and lakes, directly and indirectly supporting the functioning of downstream ecosystems and the delivery of ecosystem services (Kaltenborn et al., 2010). In fact, while being one of the most exploited resources, water is also crucial to secure the availability of other resources, such as food and energy (Clason et al., 2022). Around the globe, 1.9 billion people, living downstream of mountain areas, rely on glacier systems (Immerzeel et al., 2020). This is because meltwater from glaciers not only supplies drinking water to communities but also enables crop irrigation and aliments hydropower as a source of renewable energy (Clason et al., 2022).

Glacier floodplains also represent important areas of biodiversity, with basins drained by glaciers hosting half of the global biodiversity hotspots and a third of terrestrial species diversity (UNESCO, 2022). In addition, glaciers are of great cultural value to many people (MRI, 2018), representing sacred places for local communities and attracting millions of tourists globally (UNESCO & IUCN, 2022). Hence, both ecosystems and livelihoods are strongly dependent on glaciers, as well as on the sediments and nutrients delivered by meltwater (Clason et al., 2022).

Throughout the year, the role of meltwater in feeding our rivers is subject to many variations. While mass is gained from precipitation in the upper portion, it is then lost by melting in the lower portions of the glacier (UNESCO & IUCN, 2022). As water supply strongly relies on rainfall, meltwater can act as a buffer during dry periods (Clason et al., 2022). However, when the mass lost is greater than that the one gained, glaciers start to retreat. With temperatures around the globe rising, the retreat of glaciers has become a common pattern resulting in glaciers having become important sentinels of climate change.

While the impacts of melting glaciers are to some extent localised, the consequences extend well beyond the immediate vicinity of the glaciers, rivers, and lakes, generating ripple effects. In fact, the consequence of glacier retreat is clear: the amount of meltwater will initially – yet temporarily – increase until its peak is reached (*peak water*), after which runoff will decrease until stopping completely, failing to uphold glaciers' buffering role (UNESCO & IUCN, 2022). The temporary rise in meltwater will increase the frequency and magnitude of glacier-related hazards, such as glacial outburst floods. Similarly, as the buffering role of glaciers is lost, drought events are expected to occur with higher frequency (Taylor et al., 2023). While several glaciers are estimated to have passed their level of "peak water" already, many others are expected to reach it soon (Rounce et al., 2023), hinting at an increasing threat of floods. Globally, 15 million people are exposed to impacts from potential glacial lake outburst floods (Taylor et al., 2023). In addition, meltwater from glaciers is contributing to global sea level rise, a phenomenon which can already be witnessed extensively for the Greenland and Antarctic ice sheets (Hugonnet et al., 2021). It is estimated that the melting of all glaciers and ice sheets would lead to an increase of global sea level by 70 metres (NSIDC, 2024).

Eventually, the retreat and disappearance of glaciers will reduce the availability of water to downstream communities, and will challenge human livelihoods, ecosystem preservation and, as a result, the economy (Haeberli & Weingartner, 2020). Aquatic habitats across glacier floodplains are altered by these changes, resulting in threats to endemic and rare species (Fell et al., 2017). In addition, tourism is dramatically impacted by climate change as glaciers become more and more difficult and dangerous to access, and in a way are also losing their attractiveness and aesthetic value (Salim et al., 2021).

In many regions around the world, the economic impacts of glacier retreat are particularly severe. This is especially the case for those activities that rely directly on ecosystem functions that may be altered by climate change and by changes to glacier systems: For instance, in Mongolia, glaciers in the high mountain ranges provide more than 70% of fresh water resources so that the drastic changes in the regional glacier system are leading to even more water shortages and desertification (Orkhonselenge & Harbor, 2018). This is particularly worrisome since the country's economy relies on pastoral animal husbandry, agriculture, and a nomadic lifestyle. The strong dependence on glacial melt for water supplies and the sensitivity of the country to climate change are thus expected to lead to detrimental socio-economic impacts, in addition to the environmental consequences (Orkhonselenge & Harbor, 2018).

Crackling sounds of moving ice and deep grey wrinkles on its surface are the inevitable signs of slowly disappearing glaciers. Between 2000 and 2019, 267 gigatonnes (Gt) of glacial ice have been lost per year, an amount equivalent to 21% of the observed sea-level rise (Hugonnet et al., 2021). Since this loss is largely attributed to anthropogenic climate change (IPCC, 2019), it is estimated that, given the continuously rising temperatures, the total mass of mountain glaciers will decrease by 25–50% by 2100, depending on which global temperature change scenario materialises (Rounce et al., 2023), and that one third of World Heritage glaciers will disappear by 2050 (UNESCO, 2022).

With the gloomy outlook of increasing temperatures, every fraction of a degree matters. The retreat of glaciers may already be well beyond the tipping point, and communities and biodiversity around the world are bound to face the consequences. Nevertheless, this remains a strong call for extensive climate mitigation strategies alongside innovative ways to manage water through nature, to preserve the ecosystems that rely on this resource, and to protect communities from inevitable risks. In other words, what is needed are immediate adaptation measures to deal with the consequences to this phenomenon.

Bibliography

Clason, C., Rangecroft, S., Owens, P. N., Łokas, E., Baccolo, G., Selmes, N., Beard, D., Kitch, J., Dextre, R. M., Morera, S., & Blake, W. (2022). Contribution of Glaciers to Water, Energy and Food Security in Mountain Regions: Current Perspectives and Future Priorities. *Annals of Glaciology*, 63(87-89), 73–78. https://doi.org/10.1017/aog.2023.14.

Fell, S. C., Carrivick, J. L., & Brown, L. E. (2017). The Multitrophic Effects of Climate Change and Glacier Retreat in Mountain Rivers. *Bioscience*, 67(10), 897–911. https://doi.org/10.1093/biosci/bix107.

Haeberli, W., & Weingartner, R. (2020). In Full Transition: Key Impacts of Vanishing Mountain Ice on Water-Security at Local to Global Scales. *Water Security*, 11, 100074. https://doi.org/10.1016/j.wasec.2020.100074.

Harvey, C. (2023). *Half of All Mountain Glaciers Are Expected to Disappear by 2100. Scientific American*. Available from: https://www.scientificamerican.com/article/half-of-all-mountain-glaciers-are-expected-to-disappear-by-2100/ [Last accessed: 26th July 2024].

Hugonnet, R., McNabb, R., Berthier, E., Menounos, B., Nuth, C., Girod, L., Farinotti, D., Huss, M., Dussaillant, I., Brun, F., & Kääb, A. (2021). Accelerated Global Glacier Mass Loss in the Early Twenty-First Century. *Nature*, 592(7856), 726–731. https://doi.org/10.1038/s41586-021-03436-z.

Immerzeel, W. W., Lutz, A. F., Andrade, M., Bahl, A., Biemans, H., Bolch, T., Hyde, S., Brumby, S., Davies, B. J., Elmore, A. C., Emmer, A., Feng, M., Fernández, A., Haritashya, U., Kargel, J. S., Koppes, M., Kraaijenbrink, P. D. A., Kulkarni, A. V., Mayewski, P. A., & Baillie, J. E. M. (2020). Importance and Vulnerability of the world's Water Towers. *Nature*, 577(7790), 364–369. https://doi.org/10.1038/s41586-019-1822-y.

IPCC (2019). Summary for Policymakers. In: H.-O. Pörtner, D. C. Roberts, V. Masson-Delmotte, P. Zhai, M. Tignor, E. Poloczanska, K. Mintenbeck, A. Alegría, M. Nicolai, A. Okem, J. Petzold, B. Rama, & N. M. Weyer (Eds.) *IPCC Special Report on the Ocean and Cryosphere in a Changing Climate* Cambridge University Press, pp. 3–35. https://doi.org/10.1017/9781009157964.001.

Kaltenborn, B. P., Nellemann, C., & Vistnes, I. I. (2010). *High Mountain Glaciers and Climate Change – Challenges to Human Livelihoods and Adaptation*. United Nations Environment. Programme. Available from: https://www.unep.org/resources/report/

high-mountain-glaciers-and-climate-change-challenges-human-livelihoods-and [Last accessed: 26th July 2024].

MRI (2018). *Mountain Glaciers: Vanishing Sources of Water & Life*. Mountain Research Institute (MRI). Available from: https://mountainresearchinitiative.org/blog/mountain-glaciers-vanishing-sources-of-water-life/

NSIDC (2024). *Glaciers: Quick Facts*. National Snow and Ice Data Center (NSIDC). Available from: https://nsidc.org/learn/parts-cryosphere/glaciers/glacier-quick-facts [Last accessed: 26th July 2024].

Orkhonselenge, A., & Harbor, J. (2018). Impacts of Modern Glacier Changes on Surface Water Resources in Western and Northern Mongolia. *Journal of Water Resource and Protection*, 10, 559–576. https://doi.org/10.4236/jwarp.2018.106031.

Rounce, D. R., Hock, R., Maussion, F., Hugonnet, R., Kochtitzky, W., Huss, M., Berthier, E., Brinkerhoff, D., Compagno, L., Copland, L., Farinotti, D., Menounos, B., & McNabb, R. W. (2023). Global Glacier Change in the 21st Century: Every Increase in Temperature Matters. *Science*, 379(6627), 78–83. https://doi.org/doi:10.1126/science.abo1324.

Salim, E., Ravanel, L., Bourdeau, P., & Deline, P. (2021). Glacier Tourism and Climate Change: Effects, Adaptations, and Perspectives in the Alps. *Regional Environmental Change*, 21(4), 120. https://doi.org/10.1007/s10113-021-01849-0.

Taylor, C., Robinson, T. R., Dunning, S., Rachel Carr, J., & Westoby, M. (2023). Glacial lake Outburst Floods Threaten Millions Globally. *Nature Communications*, 14(1), 487. https://doi.org/10.1038/s41467-023-36033-x.

UNESCO (2022). *World Heritage Glaciers: Sentinels of Climate Change*. United Nations Educational, Scientific and Cultural Organisation (UNESCO).

UNESCO, & IUCN (2022). *World Heritage Glaciers: Sentinels of Climate Change*. United Nations Educational, Scientific and Cultural Organisation (UNESCO) and International Union for Conservation of Nature (IUCN). Available from: https://portals.iucn.org/library/sites/library/files/documents/2022-040-En.pdf [Last accessed: 26th July 2024].

USGS (2019). *Ice, Snow, and Glaciers and the Water Cycle*. US Geological Survey (USGS). Available from: https://www.usgs.gov/special-topics/water-science-school/science/ice-snow-and-glaciers-and-water-cycle?_ga=2.245698176.580559828.1711275691-1361622058.1710085474 [Last accessed: 26th July 2024].

9 The coral planter

> When I was a kid, I remember older people telling me that the ocean surrounding us used to look way more beautiful. The same people used to say that the water was populated by so many fish that it was even possible to catch a swordfish from the shore. Initially, I could not believe the picture they were painting. Only as I grew older, I started to understand what they meant: the beauty of the ocean was fading away in front of my eyes.

Kinjo is sitting on a chair in the middle of what looks like a magic place: a large corridor, filled with deep water tanks, each hosting a colourful range of corals, fish, and other plants. Over the tanks, rocky constructions are populated by algae, grass, and a wide variety of flowers. These are the headquarters of Sea Seed, an organisation based in Yomitan, central Okinawa, which plays a key role in repopulating the ocean with living corals.

Koji Kinjo was born in Okinawa in the seventies, as an only child, and raised by his mother. As a kid, he was already passionate about animals, and particularly fishes, as he was lucky enough to be able to see them swim across beautiful corals in the beach next to his family house. That is how he filled his childhood afternoons after school.

As he grew older, he started different businesses on the island. At a very young age, he already owned five restaurants and shops. However, his passion for the beauty of the sea did not vanish, and instead made space in his life: in every business he owned, he built large aquariums where he would farm corals and host all kinds of fishes. "During that time I thought I had it all: I was doing well, and thriving in my businesses. Back then, I thought that owning lots of shops and gaining prestige was the way to become a successful man".

> It was actually my mother who planted the seed for a different vision. One afternoon, as we were having a discussion, she told me how being successful isn't about whether you're rich or poor, it's about how you live and what you leave behind. For her, that was the true measure of a man. These words impressed me, but I only understood them later in life.

DOI: 10.4324/9781003493662-10

What really changed things for Kinjo was an out-of-the-ordinary summer, in 1998. During that summer, El Niño had spread all across the Japanese island, bringing water to record warm temperatures. He says, his voice trembling as the image of bleached corals appears vivid in his head,

> At that time, I was very busy running my café on the beach in Okinawa. That summer was extremely warm, it was even too hot to walk around. One early summer morning, I looked at the ocean and realised that something had changed: all corals had gone white.

"I was, and still am, shocked. For days, it was all I could think about. In only a short while, we had lost over 90% of our coral reef".

What happened was a wake-up call. He did not hesitate: in the months to come, he decided to sell some of his businesses and start carrying out research to better understand what was happening. He then started going to the ocean to examine corals firsthand. He smiles:

> Something immediately came to my mind: if corals could be farmed in my aquariums, why couldn't we try to grow them, and ideally plant them back into the sea? As I developed this idea in my head, there was no going back: it was as if it had captured all my entire being, all my interest and attention.

Kinjo's starts frowning when he talks about the very early days:

> At first, I wanted to plant corals as a trial, to test what I had in mind. But it was certainly not an easy moment in my life: to dedicate my whole time to planting coral, I was ragging up a lot of debt. There was a moment when I had 80 million yen in debt, and I was extremely desperate. Luckily, I was not alone, my wife was next to me, and said she was ready to sell the house if this was needed: she did not want me to give up on my work. Her words really touched me.

Kinjo's work, however, was attracting a lot of scepticism from scientists and coral experts who did not believe farmed coral could survive in the ocean. For them, touching nature was a taboo. "In my mind, I did not want to modify nature, but rather to improve the environment that people themselves had destroyed".

And finally came the days when Kinjo managed not only to grow corals, but also to plant them back into the sea.

> I had done so many trials, I cannot even remember. Deep down, I knew there was a possibility for them to survive in the sea, but no one believed me. And one very ordinary day, as I stood there, waist-deep in the water,

> I saw them: the corals – *my* corals – were releasing eggs, like tiny pearls rising from the branches, drifting up through the water, carried by the currents of the ocean. This is when I realised I had accomplished what I still think is a little miracle.

That was the moment that changed Kinjo's life as his activities attracted a lot of media attention. Finally, Kinjo could raise some funds and even hire more people to support his mission.

> Still, I felt that it was not enough, what was needed was a cultural shift among the local population, one that could lead them to truly value the beauty of the corals but also their role, and make them aware of how important it was to protect them.

"One first challenge that I encountered were the fishermen," continues Kinjo.

> They were often damaging the corals by throwing nets across the reef. At the same time, they slowly realised that, following my coral transplant, the number of fishes was increasing. I explained to them how corals were so crucial to fish populations in the sea, and once they saw what was in it for them, they started to change their practices, trying to throw the fishing nets in a way that wouldn't harm the corals.

Today, Kinjo's work is still really hard, but the reward it gives and the people it involves have also made it fun. Planting corals in itself is perhaps no longer the most challenging part: for Kinjo, this is rather convincing the local government and many people that surround him about the added value of his business.

> I work a lot, often all day long. And there are still many skeptics around me who do not believe my work makes sense. This sometimes makes me lose hope. But then, sometimes at night, when I turn off the light, I see my whole office shining with small plankton in the tanks, like stars in the sky, and remember all of the sudden why I started this.

Kinjo walks around the coloured tanks of his office, pointing to a few of them.

> You see these? Maybe you cannot understand the difference only with your eyes. But here, in these tanks, is where I am putting corals under stress, with higher temperatures and light. My idea is to get them stronger and stronger, so they can survive the increasingly warm temperatures of the ocean.

Recently, Kinjo noticed a species of tiny coral which was growing on deceased branches, and manages to resist water temperatures above 31 degrees. He quickly decided to take home a section of such coral, and farm it, and gave some samples to scientists and international organisations. "We need to adapt

Figure 9.1 Koji Kinjo, a coral farmer, in front of the sea.

to a changing environment around us. The corals that existed ten, fifteen years ago cannot survive anymore in the ocean of today".

His aim is to create an entire reef that can adapt to climate change and warm waters, and increasing pollution.

> What brings me hope is that, after having worked on this alone for 26 years, I finally see people replicating a similar work all around the world. Right now, I am giving expert guidance to coral farmers in Indonesia, Taiwan and China.

However, Kinjo also wants people to be realistic. He believes there are still too few people who truly care about the environment that surrounds them. His core message is linked to a pragmatic realization that we cannot go back to the past: we need to adapt to a new future.

> Adaptability is perhaps all I know and believe in. While the situation is very dire, I have hope in the adaptability of nature to hardship. Corals, for instance, are extremely adaptable to the situations they find themselves in. This is why we need to fight to make our world adaptable, while preserving its beauty.

His mum's words about the true nature of a successful man now resonate in Kinjo's ears as, after all these efforts, he finally recognises how much of a legacy he is leaving to the rest of the world, and how important the work he is carrying out will be – not only for nature, but also for mankind.

Fading corals: A treasure at stake

Stretches of colourful corals underneath the water appear to the human eye almost as an optical illusion, a game of lights and beauty just a few metres below sea level, where waves break. Yet, this is no longer a stable treasure, but one that is gradually fading away. Colour, a clear sign of life, is the first thing to disappear. This phenomenon, known as coral bleaching, has now become commonplace. In fact, behind their beauty, corals hide an unwanted truth: climate change, coupled with human activities, are destroying this important ecosystem and altering the livelihoods of those depending on it.

While often regarded as plants or rocks, corals are actually invertebrate animals. Coral reefs are populated with more than 800 species of corals (Rivera et al., 2020); they provide the habitat for life underwater to thrive. Their diverse nature is what has led them to gain the title of *rainforests of the sea*. In fact, coral reefs host more biodiversity than any other ecosystem of the world (IUCN, 2017). While occupying less than 1% of the ocean floor, spread over 100 countries and territories, they are home to over one quarter of all marine life (Souter et al., 2021), and to an estimated 830,000 species of organisms

(Fisher et al., 2015). Coral reefs also act as both shelter and natural corridors, allowing organisms to seek protection, breed, and migrate between ecosystems (Penniman, 2021).

In addition to representing hotspots for biodiversity, coral reefs provide a series of ecosystem services, supporting communities, economies, and infrastructure. With more than 330 million people worldwide living in the direct proximity of coral reefs (within 30 km) (Burke et al., 2012; Resourcewatch, n.d.) and 500 million known to directly depend on them worldwide (IUCN, 2017), this ecosystem is considered crucial for wellbeing, subsistence, and livelihoods. Just over 1% of them are coral reef fishermen who rely on this ecosystem for food as well as employment and generating a coral reef fisheries value of $6 billion (Eddy et al., 2021). But overall, the social, cultural, and economic value of coral reefs is much larger, estimated at $1 trillion (Costanza et al., 2014).

Coral reefs are widely recognised for their role in protecting coastlines from storms, hurricanes and sea level rise by reducing the energy of the waves (Burke & Spalding, 2022). It is estimated that reefs can in fact weaken waves by 97% (Ferrario et al., 2014). This becomes crucial for the people living in proximity to coastal areas: around 5.3 million people (Burke & Spalding, 2022). The important function of reducing wave energy is provided by reefs to about one quarter of coastlines of the countries that host them, for a total of approximately 230,000 km (Burke & Spalding, 2022).

It is not hard to deduce the enormous economic role played by corals in preventing damages during usual storm: while normally the total value is estimated to be around $4 billion, with more extreme events this amount increases drastically, reaching $36 billion worth of damages for 1-in-25-year level storms, and even $130 billion for 1-in-a-100-year level storms (Rivera et al., 2020).

The beauty of coral reefs attracts people from all around the world, making cultural ecosystem services such as recreation and tourism particularly important, especially in the areas where these activities are most relevant to the local economy (Fezzi et al., 2023). Coral reef tourism, related to in-water activities such as snorkelling and scuba diving, is estimated to be worth $36 billion every year and is found to be influenced by a series of factors, among which the health and state of the reef and its biodiversity stand out (Spalding et al., 2017), This hints at a direct dependency of the tourism sector on the ecosystem's health while also stressing the importance of making coral reef-related tourism as sustainable as possible to avoid detrimental impacts.

Finally, corals are also associated with cultural and spiritual values. In fact, the importance of corals finds its roots in indigenous cultures and communities who have historically cherished and protected this ecosystem for its role and resources. Today, indigenous groups still represent important stewarts of this fragile ecosystem, with nearly 40% of these groups living in island territories or close to coastal oceans (Ford et al., 2020). The seafood consumption of these coastal indigenous communities is also found to be considerably higher than other populations (Eddy et al., 2021).

Despite these positive features that make them so relevant to mankind and biodiversity, corals are under growing threats and their survival is at stake. On the one hand, human activities such as overfishing and coastal development as well as pollution are to blame. As a result of these activities, water quality is impoverished, affecting the ability of corals to adapt, grow, and even survive (Heron et al., 2017). However, there is another enemy in the picture which is considered the main threat to corals and which has been drastically changing life below water: climatic changes leading to rising water temperatures and ocean acidification (WWF, 2004). Generally, most species of corals cannot sustain temperatures above 29 degrees Celsius (Schoepf et al., 2015), while the rising acidity of the oceans make it harder for corals to build their skeleton (USNSF, 2018). When water temperatures increase by 1–2 degrees and the rise is sustained for some weeks, bleaching of corals occurs: the algae living in the corals tissues is expelled, making them turn white. This phenomenon often culminates in coral death (IUCN, 2017).

Since the 1950s, the area covered by living corals globally has halved. With the loss of coral, their biodiversity in reefs has also declined by 63% and coral-reef fish have decreased by more than half (Eddy et al., 2021). Even in a relatively short period of time, between 2009 and 2018, a progressive loss of 14% of corals has been witnessed globally, representing an area equivalent to all of Australia's coral reefs (Souter et al., 2021).

Science speaks clearly on this matter; if things are left unchanged and no action is made to reduce the threats to corals, this ecosystem will soon cease to exist. Under this scenario, all coral reefs in the 29 World Heritage sites containing them are predicted to disappear by the end of the century (Heron et al., 2017). But the picture is gloomy also considering global temperature rise to be kept at 1.5 degrees, with scientists predicting that 90% of coral reefs might still disappear by 2050 (Hoegh-Guldberg et al., 2018). Further, cyclones, hurricanes, and typhoons also represent a threat to corals, especially if their frequency and intensity increases due to climate change. In fact, powerful cyclones are found to be able to harm coral reefs even from 1,000 km away from their paths (Puotinen et al., 2020).

While globally appreciated for their aesthetic value, the ripple effect of the loss of corals is not understood by many. But it is clear that corals have much more to offer than a visual and recreational experience. Once their full functions and services are laid out, corals suddenly become much more valuable and their loss a real concern. With the loss of corals comes also a financial loss. An estimate indicates a yearly cost of $500 billion due to climate-related loss of reef ecosystem services (Nagelkerken, 2015).

The consequences of disappearing corals are perceived as disastrous only by few, but evidence is showing that a large portion of the world's population is already affected by their loss. It is therefore the fragility of this ecosystem, considered the marine ecosystem most vulnerable to climate change, that is making us increasingly aware of what is at stake. This calls for efforts to both reduce the threats and increase the resilience of this ecosystem.

In spite of the threats and the bleak figures, corals have also shown incredible resilience and have recovered in some instances from storms and cyclones as well as from some bleaching and mass mortality events. Coral restoration also offers important opportunities for selecting more resistant species of corals, and replanting efforts have been multiplying all over the world. But will there be enough time for such recovery going forward? One thing is sure, all efforts are needed to help this ecosystem turn from fragile to stable treasure, reverting this way the trend of fading corals.

Bibliography

Burke, L., Reytar, K., Spalding, M., & Perry, A. (2012). *Reefs at Risk Revisited - Summary for Decision Makers*. World Resource Institute. Available from: https://sustainabledevelopment.un.org/content/documents/1809Reefs_Summary_low.pdf [Last accessed: 18th January 2025].

Burke, L., & Spalding, M. (2022). Shoreline Protection by the World's Coral Reefs: Mapping the Benefits to People, Assets, and Infrastructure. *Marine Policy*, 146, 105311. https://doi.org/10.1016/j.marpol.2022.105311.

Costanza, R., de Groot, R., Sutton, P., van der Ploeg, S., Anderson, S. J., Kubiszewski, I., Farber, S., & Turner, R. K. (2014). Changes in the Global Value of Ecosystem Services. *Global Environmental Change*, 26, 152–158. https://doi.org/10.1016/j.gloenvcha.2014.04.002.

Eddy, T. D., Lam, V. W. Y., Reygondeau, G., Cisneros-Montemayor, A. M., Greer, K., Palomares, M. L. D., Bruno, J. F., Ota, Y., & Cheung, W. W. L. (2021). Global Decline in Capacity of Coral Reefs to Provide Ecosystem Services. *One Earth*, 4(9), 1278–1285. https://doi.org/10.1016/j.oneear.2021.08.016.

Ferrario, F., Beck, M. W., Storlazzi, C. D., Micheli, F., Shepard, C. C., & Airoldi, L. (2014). The Effectiveness of Coral Reefs for Coastal Hazard Risk Reduction and Adaptation. *Nature Communications*, 5(1), 3794. https://doi.org/10.1038/ncomms4794.

Fezzi, C., Ford, D. J., & Oleson, K. L. L. (2023). The Economic Value of Coral Reefs: Climate Change Impacts and Spatial Targeting of Restoration Measures. *Ecological Economics*, 203, 107628. https://doi.org/10.1016/j.ecolecon.2022.107628.

Fisher, R., O'Leary, R. A., Low-Choy, S., Mengersen, K., Knowlton, N., Brainard, R. E., & Caley, M. J. (2015). Species Richness on Coral Reefs and the Pursuit of Convergent Global Estimates. *Current Biology*, 25(4), 500–505. https://doi.org/10.1016/j.cub.2014.12.022.

Ford, J. D., King, N., Galappaththi, E. K., Pearce, T., McDowell, G., & Harper, S. L. (2020). The Resilience of Indigenous Peoples to Environmental Change. *One Earth*, 2(6), 532–543. https://doi.org/10.1016/j.oneear.2020.05.014.

Heron, S. F., Eakin, M., & Douvere, F. (2017). *Impacts of Climate Change on World Heritage Coral Reefs: A First Global Scientific Assessment*. UNESCO World Heritage Centre.

Hoegh-Guldberg, O., Jacob, D., Taylor, M., Bindi, M., Brown, S., Camilloni, I., Diedhiou, A., Djalante, R., Ebi, K. L., Engelbrecht, F., Guiot, J., Hijioka, Y., Mehrotra,

S., Payne, A., Seneviratne, S. I., Thomas, A., Warren, R., & Zhou, G. (2018). *Impacts of 1.5C Global Warming on Natural and Human Systems*. IPCC.

IUCN (2017). *Coral Reefs and Climate Change*. International Union for Conservation of Nature. Available from: https://iucn.org/sites/default/files/2022-04/coral_reefs_and_climate_change_issues_brief_final.pdf [Last accessed: 18th January 2025].

Nagelkerken, I. (2015). *How Will Marine Life Adapt to Warmer Oceans?* Economic Forum (WEF). Available from: https://www.weforum.org/stories/2015/10/how-will-marine-life-adapt-to-warmer-oceans/ [Last accessed: 18th January 2025].

Penniman, T. (2021). *Coral Reefs: Strategies for Ecosystems on the Edge*. International Institute for Sustainable Development. Available from: https://www.iisd.org/system/files/2021-05/still-one-earth-coral-reefs.pdf [Last accessed: 18th January 2025].

Puotinen, M., Drost, E., Lowe, R., Depczynski, M., Radford, B., Heyward, A., & Gilmour, J. (2020). Towards Modelling the Future Risk of Cyclone Wave Damage to the world's Coral Reefs. *Global Change Biology*, 26(8), 4302–4315. https://doi.org/10.1111/gcb.15136.

Resourcewatch. (n.d.). *Coral Reefs: Status and Outlook of Coral Reefs Around the World*. Available from: https://resourcewatch.org/dashboards/coral-reefs [Last accessed: 18th January].

Rivera, H., Chan, A., & Luu, V. (2020). Coral Reefs Are Critical for Our Food Supply, Tourism, and Ocean Health. We can Protect Them from Climate Change. *MIT Science Policy Review*, 1, 18–33. https://doi.org/10.38105/spr.7vn798jnsk.

Schoepf, V., Stat, M., Falter, J. L., & McCulloch, M. T. (2015). Limits to the Thermal Tolerance of Corals Adapted to a Highly Fluctuating, Naturally Extreme Temperature Environment. *Scientific Reports*, 5(1), 17639. https://doi.org/10.1038/srep17639.

Souter, D., Planes, S., Wicquart, J., Logan, M., Obura, D., & Staub, F. (2021). *Status of Coral Reefs of the World: 2020 - Summary for Policymakers*. GCRMN, ICRI, Australian Government, Australian Institute of Marine Science. Available from: https://gcrmn.net/wp-content/uploads/2022/05/Status-of-Coral-Reefs-of-the-World-2020-Summary-for-Policymakers.pdf [Last accessed: 18th January 2025].

Spalding, M., Burke, L., Wood, S. A., Ashpole, J., Hutchison, J., & zu Ermgassen, P. (2017). Mapping the Global Value and Distribution of Coral Reef Tourism. *Marine Policy*, 82, 104–113. https://doi.org/10.1016/j.marpol.2017.05.014.

USNSF (2018). *Scientists Pinpoint How Ocean Acidification Weakens Coral Skeletons*. US National Science Foundation. Available from: https://new.nsf.gov/news/scientists-pinpoint-how-ocean-acidification [Last accessed: 18th January].

WWF (2004). *Climate Change Top Threat to World's Coral Reefs*. Available from: https://www.wwf.mg/en/?16973/Climate-change-top-threat-to-worlds-coral-reefs [Last accessed: 18th January 2025].

10 Back to the roots of vines

On the very west coast of Sicily, near the town of Menfi, rolling hills are covered by sun-drenched vineyards, where the salty air is a reminder of the influence of the surrounding Mediterranean Sea. The entrepreneur and wine maker Marilena Barbera is sitting on a tall rock among the rows of vines, her gaze fixed on a point towards the horizon. The sun has just started rising, but it's already burning on the skin, despite the early morning hours.

Turning towards the rows of vineyards, Marilena takes a deep breath. All around her, the blue of the Mediterranean Sea is standing against the green of the Marsala vines covered hills and producing a very light breeze. She walks through the vineyards and starts telling her story.

"It's crazy to think all of this started from a wedding gift that my grandfather received in the 1920s: this very land".

It was Marilena's grandfather who had the idea that started it all: to plant the vineyards on his land in order to sell the grapes to the producers of Marsala wine. Her dad took over around 30 years later with a similar idea in mind, to continue selling grapes while producing a very small quantity of wine to be used by family and friends. "Back then", she explains, "they did not think about making wine as a living, as it was – and it is actually still – not an easy practice". But things started to change not long after that, when the idea of winemaking as a core business arose in Marilena's dad, as in many other farmers on the island.

For a long time, Marilena had stepped away from that world that was so familiar to her, yet unordinary. While her father was starting his own grape production, Marilena had a whole different future in front of her. "I knew I wanted to see the world and leave Sicily," she smiles, "and that's why I moved to study International Relations in Florence, thinking about a diplomatic career, and later studying to become a tax accountant. I was exploring different paths. And I had no idea I would end up back here".

Today, Marilena is an independent winemaker, although, as she loves to repeat, this was hardly her choice as "it is the land that chooses you and not the other way around". In the early 2000s, Marilena had a calling:

> I knew that my father had started reflecting upon producing wine as a core business and not as a side hustle anymore, using our grapes instead of

DOI: 10.4324/9781003493662-11

> selling them. And I felt that it was the right time to join him and go back to my hometown.

As she wanted to be useful right away, she took the role that was closest to her professional background, managing the sales of the family wine. "Little did I know that I was going to be faced with a whole company to bring forward when my dad passed away in 2006", she continues.

> It was not an easy time. My mum was a retired teacher and had little to do with the business as well. But we told each other that it was worth the risk. To do this, we needed to start studying again.

Marilena's lack of skills in the field turned out to become a blessing in disguise. "Not having a background can make you more free about the road you'd like to take," she smiles.

> You are not shaped by other people's points of view and stereotypes, which is what made my choices more natural. I soon realised that we were working in direct contact with nature, something which requires a lot of patience, and the ability to be aware of the world around you.

She started reading a lot about wine, and different forms of wine production. Very soon, she then realised how much climate change, paired with intensive agricultural practices, were affecting the industry, and could not help but wonder what winemakers themselves could do to change this.

"Only when working in close contact with the land, one realises how much damage has been done to it, and very often it is too late to do anything about it", Marilena admits. She explains how everything that the plant produces starts from its relationship with the land: the nutrients in the soil, the water in the ground, the effect of the temperature, and the inputs from the surrounding nature. And even if this seems to be an obvious concept, many people end up forgetting it, thinking that they could change the quality of the plants from one day to the other with the use of pesticides and other chemicals.

"What I quickly found out from my research and experience was that intensive agriculture, in the past forty years, has completely destroyed our lands' superficial layer". Heavy treatment of the soil with chemicals has a direct impact on the decline of beneficial soil microorganism, as well as on the quality of ground and surface water, hence on biodiversity and on human health. This is paired up with the devastating effects that climate change is having on wine production in the area.

> When I came back to Sicily, the changes in our lands and climate appeared to me as sudden and drastic. We used to say that the Mediterranean was the cradle of agriculture, as it allowed for the climatic conditions to

> develop a whole civilisation. But seasons are not stable anymore, nor are climatic conditions. And we know it will only get worse in the future.

Marilena suddenly stops talking, the eyes fixed on the vineyard. She turns and reaches to a bunch of grapes, picking it up. "The harvest is approaching and look at what I have". Holding the bunch of grapes on the palm of her hand, she starts turning it, revealing another side to it. While in the front grapes are green and perfectly shaped, in the back all grapes look like dark little raisins, less than half the size, wrinkly and dry. "This side has been completely burnt by the sun," she explains. Temperatures in Sicily have sky-rocketed the last summer of 2023, reaching 49 degrees Celsius, and precipitation has been almost non-existing for months.

For these reasons, Marilena came to the realisation that there was no chance to keep using the same technique as her father and, before him, her grandfather to cultivate the land, as many other generations of farmers had done. She started reorienting all her activity towards the preservation of biodiversity, taking inspiration from the concept of biodynamic agriculture, an ecological farming system based on living solutions, and leaving space for biodiversity to thrive in the farm. This includes avoiding synthetic pesticides and fertilisers or any other systemic farm chemicals.

> Most of the vineyard operations are performed manually, including pruning, canopy management, green harvest and defoliation. I also try to avoid the use of sulphur and copper, and prefer compost to these, or algae and vegetal teas. I also pick a small quantity of grapes for them to undergo spontaneous fermentation before the harvest to generate strong enough yeast. To add acidity to the wine, I normally use a small quantity of unripe grapes, which add natural levels of acids to the wines during their fermentation.

"There's nothing chemical herbicides can do to avoid the sun from burning these grapes", she exclaims.

> What I rather try to do, is to create conditions for biodiversity to increase: not only do I not want to pollute the soil, but I also want to enable the soil to be more fertile and grow other plants. These plants, in turn, could offer habitats for biodiversity and bring shadow to my vineyards, thereby preserving them from being burnt or unnourished.

"I get so frustrated when I see winemakers, such as my neighbour, thinking that chemicals would solve all their problems," she says. Spraying very powerful herbicides on the vineyard may do the trick in the short term, but inevitably has the effect of destroying all the microorganism in the soil in the long term. The unexpected consequence is that the vineyard is not able to produce sufficient quantities for its sustenance, drying out almost completely.

Marilena is committed to continue to change things through the use of natural and biodynamic practices and wants to encourage such change in her fellow farmers, stressing the idea that the solution can only lay in nature itself. Agriculture, of any kind, is so reliant on nature that it is hard to ignore the way climate is affecting it. Despite not being a necessary resource, wine has always been a product of great cultural and social value, but the production of such a resource

Figure 10.1 Marilena Barbera, sustainable wine grower in Sicily, pictured in her vineyard.

needs to go hand-in-hand with the factors that make it happen. And wine is also just an example of how these practices could change the way agriculture is resilient to climate change worldwide. This is a challenge and a mission for farmers around the world, as well as an opportunity for those who will embrace the practice in the future as new wine regions emerge.

Climate sensitive crops: Rethinking agriculture with and for nature

The concepts of "agriculture" and "sustainability" do not always go hand-in-hand. The challenge in combining them arises from the realisation of both the impacts and dependencies of agriculture on nature, as is the case for many other economic activities. Agriculture is known as a key contributor to the biodiversity and climate crises, being responsible for environmental degradation, and in particular for one-third of global greenhouse gas emissions due to production processes and changes in land use (Ritchie, 2019; Semov, 2021), with animal production as the main contributor (Agudelo Higuita et al., 2023). At the same time, it is a fundamental human activity and one of the sectors most reliant on natural processes and therefore strongly impacted by climate change. This paradox sheds light on the need to alleviate the negative relationship between economic activities such as agriculture and nature, and has led to the rise of many agricultural approaches which refer to the sustainability component of the practice (Oberč & Schnell, 2020). These practices reflect the urgency to rethink agriculture to consider such a global industry from a resilience and sustainability angle.

While more than one-third of our land is used for agriculture (FAO, 2020), over half of global agricultural land is already degraded (Noel & Stewart, 2015), a figure surpassing 80% in the European Union (JRC, 2024). The direct consequence of such degradation is reduced productivity, with losses estimated at $400 billion a year (Masterton, 2022) and biodiversity loss (JRC, 2024). As a matter of fact, agriculture is causing more harm to biodiversity, natural resources, and soil than any other industry (Miralles-Wilhelm & Iseman, 2021). This is mainly a result of an intensive farming approach which involves the use of heavy machinery, pesticides, and fertilisers, as well as the release of microplastics in the soil (Nizzetto et al., 2016). As a consequence, agriculture has become the main driver of the loss of 80% of natural habitats (Miralles-Wilhelm & Iseman, 2021) and the sector that has the highest water consumption while simultaneously contributing the most to water pollution (OECD, 2017). Intensive farming also has serious health consequences to both farmers that are exposed to a lot of pesticides, as well as the surrounding communities through the wider air and water pollution caused by the production process and leading to potential poisoning and illnesses (Douglas et al., 2018; Health Care Without Harm, 2018; ILO, 2011).

At the same time, agriculture is also particularly vulnerable to climatic changes (IPCC, 2014) and the impacts on agriculture are becoming increasingly obvious. As lands, plants, and soils are the foundation of agriculture, climatic changes affecting their status will inevitably lead to alterations in the sector's functionality. For example, variations in rainfall and temperature can interfere with the growth and maturing process of crops, while increased temperature can also expand the range of pests and diseases for plants. Similarly, the structure of the soil may also be affected, making it less fertile and reducing how much crops it can produce (Bibi & Rahman, 2023). And this trend is hardly reversible: in particular, soil erosion is considered the biggest threat to soil fertility. In Europe, 12 million hectares of land suffer from soil erosion and the detrimental impacts on productivity are translated into an economic loss to farmers of €1.25 billion annually (Panagos et al., 2018). In addition, water, which is a key resource for agriculture, is under increasing threat due to climate change. Almost half of cultivated systems are located in drylands (Noel & Stewart, 2015) and the majority of all irrigated farms worldwide are subject to water shortages (Richter, 2016). In regions already considered water-scarce, 90% of water consumption is related to irrigation for agriculture (Richter, 2016). Unsurprisingly, it is expected that the situation will only worsen, meaning that by 2050, agricultural water scarcity will become more severe in 80% of croplands (Liu et al., 2022).

Facing this looming scenario, it is clear how the sustainability of agriculture is under threat, posing an increasing risk to food security, health, biodiversity, and resilience around the world. Farmers producing sensitive crops are experiencing the impacts of climate change first-hand, and as the on the ground stewards of lands around the world, they are faced with an important responsibility and a difficult mission to deviate from traditional practices that may have been in place for centuries in order to transition to sustainable agriculture.

Vines, while representing a relatively small fraction of the agriculture sector and occupying a rather small portion of land, are still associated with important negative environmental impacts, especially if traditional, intensive farming practices are applied (Bandinelli et al., 2020; Engstrom, 2023). At the same time, they are an important example of the changes happening to agriculture, as they are among the most weather-sensitive agricultural crops (Ashenfelter & Storchmann, 2016). Because of their reliance on climate for crop location, quantity, taste, and health, vines are strongly linked to, and hence dependent on, the status of soils, water, and temperature. As a matter of fact, an increase in temperature can alter the geographical distribution of viticulture (Biasi et al., 2019; Droulia & Charalampopoulos, 2021). More precisely, an increase by 2 degrees Celsius would reduce the land suitable for wine production by more than 50% of (Petriashvili et al., 2023). Adaptation to such changes requires change in grape varieties to those more resistant to climatic changes, as well as changes in the cultivation systems.

In Europe, grapevines are of great importance due to their social, cultural, and economic role. In fact, Europe is the continent with the largest wine production, especially in the Mediterranean region (Droulia & Charalampopoulos, 2021). However, wine geography is bound to change. In fact, exposure to very high temperatures, above 35 degrees Celsius, negatively impacts the crops, causing the grapes to burn or ripe. At the same time, temperatures lower than usual and more frequent spring frosts damage the grapevine development, affecting flowering and harvesting times (Droulia & Charalampopoulos, 2021; Pastore et al., 2022). Similarly, changes in precipitation have proved particularly dangerous to vines with, on the one hand, low precipitation leading to drought events and causing grapevines to get sunburnt, and, on the other hand, high precipitation leading to floods and the related inability of the soil to absorb water, resulting in drowned vines (Droulia & Charalampopoulos, 2021). All these climate variations have impaired production activities around the world, and particularly in countries with peak heatwaves. While the suitability of the current wine regions will be increasingly endangered, new suitable areas might arise in the future (e.g. in central, northern, and western Europe) (Moriondo et al., 2013).

There are ways to limit the progressive degradation of wine production, and these include approaches to sustainable agriculture, such as nature-inclusive,[1] regenerative,[2] or biodynamic agriculture.[3] Even if limited compared to conventional agriculture, these approaches are being increasingly adopted as a means to adapt agricultural systems to climate change. These techniques are based on sustainable practices giving space to and taking care of biodiversity and ensuring positive impacts on the environment. While they are all different in nature, they share a mission to support agricultural productivity while enhancing natural processes, striving for sustainability in all its considerations (environmental, social, and economic). It is clear that the priority of preserving the environment lives alongside that of providing safe and healthy food for all (Oberč & Schnell, 2020), and these are often in opposition. For this reason, now more than ever, there is a need to explore and expand these agricultural practices more widely, turning them from mere unconventional innovations into the new normal.

Notes

1 Nature-based agriculture is an economically viable agriculture system that optimally manages natural resources and provides a basis for sustainable business operations, including caring for ecological functions and biodiversity on or around the business, that is, farm (van Doorn et al., 2016)

2 Regenerative agriculture consist of practices that regenerate soil, reducing but not necessarily eliminating synthetic pesticides and fertilisers, and going beyond the reduction of negative effects towards ensuring that agriculture has a positive effect on the environment (FOLU, 2019)

3 Biodynamic agriculture is an ecological farming system that views the farm as a self-contained and self-sustaining organism (Demeter, 2012; Kristiansen & Merfield, 2006).

Bibliography

Agudelo Higuita, N. I., LaRocque, R., & McGushin, A. (2023). Climate Change, Industrial Animal Agriculture, and the Role of Physicians – Time to Act. *The Journal of Climate Change and Health*, 13, 100260. https://doi.org/10.1016/j.joclim.2023.100260.

Ashenfelter, O., & Storchmann, K. (2016). The Economics of Wine, Weather, and Climate Change. *Review of Environmental Economics and Policy*, 10(1), 25–46. https://doi.org/10.1093/reep/rev018.

Bandinelli, R., Acuti, D., Fani, V., Bindi, B., & Aiello, G. (2020). Environmental Practices in the Wine Industry: An Overview of the Italian Market. *British Food Journal*, 122(5), 1625–1646. https://doi.org/10.1108/BFJ-08-2019-0653.

Biasi, R., Brunori, E., Ferrara, C., & Salvati, L. (2019). Assessing Impacts of Climate Change on Phenology and Quality Traits of *Vitis vinifera L.*: The Contribution of Local Knowledge. *Plants*, 8(5), 121. https://www.mdpi.com/2223-7747/8/5/121

Bibi, F., & Rahman, A. (2023). An Overview of Climate Change Impacts on Agriculture and Their Mitigation Strategies. *Agriculture*, 13(8), 1508. https://www.mdpi.com/2077-0472/13/8/1508

Douglas, P., Robertson, S., Gay, R., Hansell, A. L., & Gant, T. W. (2018). A Systematic Review of the Public Health Risks of Bioaerosols from Intensive Farming. *International Journal of Hygiene and Environmental Health*, 221(2), 134–173. https://doi.org/10.1016/j.ijheh.2017.10.019

Droulia, F., & Charalampopoulos, I. (2021). Future Climate Change Impacts on European Viticulture: A Review on Recent Scientific Advances. *Atmosphere*, 12(4), 495. https://www.mdpi.com/2073-4433/12/4/495

Engstrom, E. (2023). Regenerative Viticulture: What Does It Mean for the Wine Industry? Available from: https://granthaminstitute.com/2023/05/31/regenerative-viticulture-what-does-it-mean-for-the-wine-industry/ [Last accessed: 18th July 2024].

FAO (2020). *The State of Food and Agriculture*. Overcoming Water Challenges in Agriculture. Food and Agriculture Organization (FAO) of the United Nations. Available from: https://doi.org/10.4060/cb1447en

FOLU (2019). *Growing Better: Ten Critical Transitions to Transform Food and Land Use*. The Food and Land Use Coalition (FOLU). Available from: https://www.foodandlandusecoalition.org/wp-content/uploads/2019/09/FOLU-GrowingBetter-GlobalReport.pdf

Health Care Without Harm. (2018). *Community health risks of industrial agriculture*. Available from: https://foodcommunitybenefit.noharm.org/resources/community-health-needs-assessment/community-health-risks-industrial-agriculture [Last accessed: 18th July 2024].

ILO (2011). *Safety and Health in Agriculture*. International Labour Organisation. Available from: https://www.ilo.org/resource/other/safety-and-health-agriculture [Last accessed: 18th July 2024].

IPCC (2014). Summary for Policymakers. In: C. B. Field, V. R. Barros, D. J. Dokken, K. J. Mach, M. D. Mastrandrea, M. C. T.E. Bilir, K. L. Ebi, Y. O. Estrada, R. C. Genova, B. Girma, E. S. Kissel, A. N. Levy, S. MacCracken, P. R. Mastrandrea, & L. L. White (Eds.) *Climate Change 2014: Impacts, Adaptation, and Vulnerability. Part A: Global and Sectoral Aspects. Contribution of Working Group II to the Fifth Assessment Report of the Intergovernmental Panel on Climate Change*. Cambridge University Press, pp. 1–32.

JRC (2024). *EUSO Soil Degradation Dashboard*. European Commission Joint Research Centre. Available from: https://esdac.jrc.ec.europa.eu/esdacviewer/euso-dashboard/ [Last accessed: 2nd September 2024].

Liu, X., Liu, W., Tang, Q., Liu, B., Wada, Y., & Yang, H. (2022). Global Agricultural Water Scarcity Assessment Incorporating Blue and Green Water Availability Under

Future Climate Change. *Earth's Future*, 10(4), e2021EF002567. https://doi.org/10.1029/2021EF002567.

Masterton, V. (2022). *What Is Regenerative Agriculture?* World Economic Forum. Available from: https://www.weforum.org/agenda/2022/10/what-is-regenerative-agriculture/

Miralles-Wilhelm, F., & Iseman, T. (2021). *Nature-Based Solutions in Agriculture: The Case and Pathway for Adoption*. Food and Agriculture Organization of the United Nations and The Nature Conservancy. Available from: https://openknowledge.fao.org/server/api/core/bitstreams/9c6d587e-1532-4252-852f-d2657634a66a/content [Last accessed: 18th July 2024].

Moriondo, M., Jones, G. V., Bois, B., Dibari, C., Ferrise, R., Trombi, G., & Bindi, M. (2013). Projected Shifts of Wine Regions in Response to Climate Change. *Climatic Change*, 119(3), 825–839. https://doi.org/10.1007/s10584-013-0739-y

Nizzetto, L., Futter, M., & Langaas, S. (2016). Are Agricultural Soils Dumps for Microplastics of Urban Origin. *Environmental Science and Technology*. https://doi.org/10.1021/acs.est.6b04140

Noel, S., & Stewart, N. (2015). *Report for Policy and Decision Makers: Reaping Economic and Environmental Benefits from Sustainable Land Management*. Economics of Land Degradation Initiative. Available from: https://www.eld-initiative.org/fileadmin/pdf/ELD-pm-report_05_web_300dpi.pdf [Last accessed: 18th July 2024].

Oberč, B. P., & Schnell, A. (2020). *Approaches to Sustainable Agriculture*. IUCN.

OECD. (2017). *Water Risk Hotspots for Agriculture*. Available from: https://doi.org/10.1787/9789264279551-en

Panagos, P., Standardi, G., Borrelli, P., Lugato, E., Montanarella, L., & Bosello, F. (2018). Cost of Agricultural Productivity Loss Due to Soil Erosion in the European Union: From Direct Cost Evaluation Approaches to the Use of Macroeconomic Models. *Land Degradation & Development*, 29(3), 471–484. https://doi.org/10.1002/ldr.2879.

Pastore, C., Frioni, T., & Diago, M. P. (2022). Editorial: Resilience of Grapevine to Climate Change: From Plant Physiology to Adaptation Strategies [Editorial]. *Frontiers in Plant Science*, 13. https://doi.org/10.3389/fpls.2022.994267

Petriashvili, A., Mach, J., Štěbeták, M., Prášilová, M., Svoboda, R., Navrátilová, M., Beranová, M., Veselá, K., Hofman, V., & Němec, O. (2023). The Impact of Climate Change on the Sustainability of Wine Production and the Structure of Its Consumption in Czechia. *Heliyon*, 9(7), e17882. https://doi.org/10.1016/j.heliyon.2023.e17882.

Richter, B. (2016). *Water Share: Using Water Markets and Impact Investment to Drive Sustainability*. The Nature Conservancy.

Ritchie, H. (2019). *Food Production Is Responsible for One-Quarter of the world's Greenhouse Gas Emissions*. Our World In Data. Available from: https://ourworldindata.org/food-ghg-emissions [Last accessed: 18th July 2024].

Semov, A. (2021). *Why Empowering 100 Million Farmers to Transform Our Food Systems Matters*. World Economic Forum. Available from: https://www.weforum.org/agenda/2021/09/why-empowering-100-million-farmers-to-transform-our-food-systems-matters/ [Last accessed: 18th July 2024].

11 The resilience of a wandering village

> When it happened, we were all asleep as it was late at night. A sudden, deep rumble jolted us awake, shaking the ground beneath our feet. As we walked outside of our houses trying to make sense of what was happening, we watched with horror as the earth from a nearby hill gave way, falling down with what seemed an unstoppable force.

Vilmaina Ratu, a 44-year-old woman from the Tukuraki village, situated in Fiji's main island of Viti Levu, tells us her direct experience of the landslide that struck her village in 2012. This was just the first of the disruptive events she experienced, as similar ones occurred over the course of the past decade, a time through which her life changed forever. On January 26, 2012, around 939 millimetres of rain fell over three days in the peaceful village in the Yakete District, causing a massive landslide which buried a family of four, along with their house. The landslide destroyed almost 80% of the village, cutting road access, water supply, wiping out crops and infrastructure. Vilmaina says:

> All of the sudden, our village was left isolated, cut off from the outside world, and our homes and livelihoods were buried beneath the weight of the earth. It took two long, heartbreaking days to retrieve the bodies of the young family. It was something that forced us to face a painful reality: we could no longer stay in the place we had always called home.

Vilmaina remembers how everything had to happen so quickly. Once the news of the landslide reached the local government office (the Commissioner's Western Office), they gave villagers only eight hours to evacuate the old village site, which was hardly enough time to prepare a proper burial for the four victims. For this reason, villagers had to find quick solutions, asking for shelters in neighboring communities.

> We were so grateful to our neighbors who took us in during that difficult time. But we knew we couldn't stay with them forever; it was hard for their families too. As a mother, living in a temporary place made it almost

DOI: 10.4324/9781003493662-12

> impossible to provide what my kids needed, especially after the landslide, and the drought that followed, which left us with no income.

At the beginning, villagers thought it was only a question of months before the village could be made safe again, and before they could return to their houses. During these months, some families had attempted to build some temporary structures next to the old village area, with the aim of providing something that could feel like home for the community. However, less than a year later, Tropical Cyclone Evan, a Category 4 storm, hit the same Tukuraki's temporary structures.

Vilmaina remembers the desperation of that time.

> Our shelters were battered, crops were uprooted, and anything we had managed to salvage was destroyed by the storm's winds and flooding. Some of us tried to return to the old site, clinging to any material left behind to rebuild temporary structures. It was difficult, but at the time we were determined to keep our community together and our village alive, no matter how fragile it was slowly becoming.

In the meantime, the Yakete District, Ba Provincial Council, and national agencies were working to find a new, safer place to relocate us permanently.

"For us, land is everything. Not having a land, a place where to build our properties again, where to practice our village rituals and to grow crops, was stripping our identity away" Vilmaina continues.

Unfortunately, the landslide and the cyclone were not the last catastrophic event hitting the village. In February 2016, just a few months after the government confirmed a new relocation site for the village, Cyclone Winston struck. This was the strongest cyclone Fiji had ever seen, a Category 5 storm with winds reaching 320 km/h.

> At the time we had started living in the temporary shelters near the old village houses – which obviously stood no chance against this force. With barely any time to gather supplies, many of us fled to the nearby caves in the Ba highlands, bringing only what we could carry to survive. We were five families sheltered in the caves, and we spent an entire month there, rationing the small amount of food we had taken. It was an experience none of us was ready for, something we will hardly forget.

Vilomena remembers getting back to the village area after the cyclone:

> The storm left nothing behind in the village – the temporary homes, our few belongings, even our crops that we depended on for food and income were gone. That was when the government declared our village a priority for relocation, recognizing the need to keep us safe from further disasters.

Experts from the Mineral Resources Department studied the land in the old village and confirmed what we already feared: the soil in the area was too loose, too unstable for long-term settlement. After a long discussion with the nearby Mataqali Yalimara tribe, they managed to find a nearby site with more secure soil.

In August 2016, the Fiji National Steering Committee, working with the Pacific Community (an international development organisation focused on the Pacific), launched a project to rebuild a new Tukuraki. The project followed the best practices to protect us from future disasters. But getting the materials to start building was yet another challenge.

Vilomena remembers

> With the widespread damage Cyclone Winston left across Fiji, resources were scarce. Yet, even through these challenges, we could finally see the end of our troubles: our real home was no longer existing, but a new stable one was finally within reach.

When asked about the old village, Vilomena stresses how easy life was.

> Our old Tukuraki village was a very rich village. When I got married and moved here, I saw that everything was easy, especially for us women, since most of us had kids, we saw that the food source was abundant. This included vegetables, and fruits, root crops, as well as plantains.

Now that they relocated, Vilomena recalls the older times with a little nostalgia, reminiscing of simpler, happier, but also richer times. Farming in Tukuraki was always organised by families, and traditionally, the women here hold the rights to the farming areas. Not having any land for farming means that, for many women, the new village doesn't offer any source of income.

> To tend our old crops, we have to walk about 10 kilometers across rugged, hilly ground, and carry heavy loads back to the new village. The trip can be especially hard for mothers like me, who bring our young children along when our husbands are away. We still need to farm for our food, but without local plots, we can no longer grow enough to sell at the market. Financially, this has been a big setback for me.

At the same time, the new Tukuraki village has given a safe home with clean water, electricity, and a school nearby for the children. Before, children as young as at kindergarten age used to walk many kilometres every day to get to school, and often needed to sleep in dormitories if they did not manage to make it home before dark.

"I cannot deny that our new village has brought us comfort in many ways, but the transition has still been challenging". One of the main issues, according to Vilomena, is that villagers were not directly included in the planning of the houses.

> During heavy rains, if a large truck drives by, we feel the house tremble slightly, and each time, this brings back memories of the landslide and the trauma of watching the ground shift beneath us. Now, we have asked that big trucks use a different route to avoid this unsettling movement.

Figure 11.1 Vilmaina Ratu, a woman from the relocated Tukuraki village.

> Life here is different in other ways, too. Even though we have electricity now, and our children have good light to study, the evenings are hotter than they used to be. It's becoming harder to rest well as the climate shifts around us. When it rains, the roads often wash out, leaving us and our neighbors to repair them ourselves. Each change in weather reminds us of what we went through. Although we are standing on safer ground now, we are still affected by climate change, day in and day out.

Moving to a new village has also brought deep changes within the new generation. Vilomena fears that many kids are starting to lose touch with the traditions that have always held our village together.

> Without land nearby, I worry that we are losing part of our cultural identity. We do our best to pass on the traditional ways to our children, but with the changes and demands of life in the new village, it feels more and more like a struggle to keep our heritage alive.

Unstable grounds: How landslides reshape lives and traditions

Sometimes, even the earth below our feet, something commonly considered strong and stable, may become fragile, lose its grip, and start to strip off, leaving us to wonder if anything can really be long-lasting. This is what happens during a landslide. As a mass movement of earth, rock, or debris down a slope, a landslide is a destabilising, and in many cases even catastrophic, event for nearby populations. Their high velocity and impact pressures make rapid landslides particularly devastating. The element of surprise leads to unexpected impacts for surrounding communities and economies. However, while these are the most common landslides we hear of, some landslides might also occur slowly, gradually developing over time, giving the impression that nothing is changing, while indeed doing so drastically (Pollock & Wartman, 2020).

Overall, landslides are more widespread than any other geological event (WHO, 2025) and are among the most critical natural disasters (Alimohammadlou et al., 2013). While gravity plays its role and soil on slopes naturally moves downhill, landslides are caused and influenced by different factors, such as temperature, precipitation and extreme weather events, seismic activity, changes to vegetation, as well as human activities (Asada & Minagawa, 2023; Gariano & Guzzetti, 2016). These factors contribute to surface erosion and runoff processes, which in turn make the ground more unstable (Gariano & Guzzetti, 2016).

While earthquakes represent a key cause of landslides, these are frequently triggered by extreme meteorological events (Tichavský et al., 2019). Water, especially, triggers mass movements. Intense rainfall, whether long lasting or not, is a key driver of landslides (Tichavský et al., 2019) as heavy rainfalls alter surface and groundwater dynamics, contributing to slope instability (Guzzetti et al., 2022). So far, on average, about 400 thousand significant landslides triggered by rainfall have occurred annually between 1980 and 2018 (World Bank, 2020). As a matter of fact, climate change has been materialising in increased frequency and intensity of rainfall events in many areas, pointing to a rising trend in extreme rainfall-generated landslides (Asada & Minagawa, 2023).

At the same time, temperature also influences the strength, viscosity, and moisture of soil (Loche & Scaringi, 2025). As we know, the earth's climate shows a warming trend (IPCC, 2021). Longer dryer conditions may affect soil by increasing desiccation, leading to cracks that destabilise the ground, and positioning temperature as a potential factor contributing to slope instability (Ganepola & Jayawardena, 2020). Such cracks allow rainwater to infiltrate the soil, leading to more groundwater levels and slope instability. Furthermore, temperature is a critical factor for vegetation which increases soil stability through roots. As such, changes in temperature inevitably lead to changes in vegetation, determining yet another impact on soil itself and therefore on the occurrence of landslides (Asada & Minagawa, 2023).

The combination of intense rainfall and high temperatures has been investigated. Research is showing that the combined effect has been linked to landslides activity (Tichavský et al., 2019). In addition, temperature is found to interact with soil evapotranspiration, potentially having a knock-on effect on precipitation patterns that trigger landslides (Tagarelli & Cotecchia, 2020).

Human activities are also influential factors in the generation of landslides. In fact, activities that change the topography as well as surface and subsurface hydrology, such as those linked to construction and agriculture, negatively affect slope stability (Tian et al., 2025). In addition, land use changes such as the transition from forested areas to bare or cultivated land are often associated with shallow landsliding, indicating that land use and land use cover management can play a role in reducing these hazards (Tian et al., 2025; Xu et al., 2021).

The results are direct and indirect impacts on communities, the environment that surrounds them, the land that hosts them, as well the economies that sustain them. Local communities living in areas prone to landslides, such as in the proximity of hills and mountains and where climatic changes represent the above mentioned features, are particularly vulnerable. People's homes and critical infrastructures, such as buildings and bridges, are directly facing the threats of landslides as the fast movement of earth, rock, or debris will hit anything it crosses down the slope (Luo et al., 2023). People's health is also impacted, both at physical – due to the strength and devastating nature of these events – and mental levels – as landslides can have dire consequences in terms of fatalities, injuries, and disruption of living situations (Climate ADAPT, 2024). Yet, what determines vulnerability goes beyond the intensity of the event in itself as it also depends on many human factors. For instance, vulnerability to landslides appears to be higher for people in economically developing countries, almost twice as much as developed countries, as is the case with many environmental disasters. The same goes for the aftermath of landslide events, with less people being rescued in economically developing countries compared to developed ones (Pollock & Wartman, 2020).

So far, landslides have affected a considerable amount of people around the world with an estimate indicating about 4.8 million people and more than 18 thousand deaths between 1998 and 2017 (WHO, 2025). The economic loss due to damages from landslides reaches $20 billion per annum (Sim et al., 2022) and in industrialised countries alone, the cost exceeds $6 billion (ISPRA, 2011). In terms of human losses, the mountains of Asia have historically been the global hotspot for landslides (Froude & Petley, 2018) where the primary driver is rainfall, associated with large-scale meteorological phenomena such as monsoons – and the latest data projects an increase in landslide hazards in this region due to climate change (Stanley et al., 2024).

At the same time, climatic, topographic, and anthropogenic features are making landslides more likely all over the world (Fidan et al., 2024). In particular, it is expected that both the frequency and magnitude of landslides will increase with climate change, even though the complexity of environmental and human features affecting these events challenges this relationship (Jemec Auflič et al., 2023). In addition, the social aspects and impacts of landslides remain understudied or underestimated in global risk databases (Froude & Petley, 2018). However, fatalities do not tell the whole story: while landslide-related mortality rate is significant, landslides tend to have repercussions well beyond those directly physically affected and prolonging in time. Post-landslides recovery is in fact considered to take a long time as it implies dealing with people's financial, occupational, psychological, and domestic problems, in addition to physical and infrastructural recovery (Pollock & Wartman, 2020).

Experiences of relocation and of communities losing their land remain, to some extent, untold stories as the emotional element behind such emergency responses is deeply rooted in personal feelings. Community relocation is acknowledged as a common emergency response, yet representing one of the most complex climate change–related governance challenges (Bronen, 2021). When communities are relocated in response to a disaster, risks of different nature arise. In particular, from a socio-cultural perspective, leaving one's land behind leads to a sense of community disintegration, people might experience what is known as loss of local memory and even their culture and traditions – normally linked to their land – might change drastically or even be lost (Ietto et al., 2022).

Around the world, there are many ghost towns and wandering villages, behind which hide deeply injured communities trying to start new. Climate change and human pressures are not only increasing the likelihood of disasters and their related consequences, but are also creating dire conditions that may worsen the adaptation capacity for when disasters such as landslides occur. Cyclones and other extreme events are an example of such conditions, as, when they occur after a disaster has already taken place, they also challenge the implementation of emergency responses such as post-disaster relocations. Once again, it is the resilience and willingness of people to deal with such changes that allows for overcoming such difficult situations. Disaster risk preparedness, awareness, and education programmes play a crucial role in reducing the number of people at risk of landslides as well as in alleviating the damages and reducing vulnerability (Shaw et al., 2009). However, access to hazard early warning systems and other recovery and risk reduction resources is often lacking, or unequally distributed (WMO & UNDRR, 2022). Increasing the adaptive capacity of both people and places to landslides and other related disasters will require looking for innovative preventative solutions as well as investing in resources to ensure recovery takes place considering all the physical, emotional, and economic tolls it comes with.

Bibliography

Alimohammadlou, Y., Najafi, A., & Yalcin, A. (2013). Landslide Process and Impacts: A Proposed Classification Method. *CATENA*, 104, 219–232. https://doi.org/10.1016/j.catena.2012.11.013.

Asada, H., & Minagawa, T. (2023). Impact of Vegetation Differences on Shallow Landslides: A Case Study in Aso, Japan. *Water*, 15(18), 3193. https://www.mdpi.com/2073-4441/15/18/3193

Bronen, R. (2021). Rights, Resilience and Community-Led Relocation: Creating a National Governance Framework. *Harbinger*, 45, 25–45. Available from: https://socialchangenyu.com/harbinger/rights-resilience-and-community-led-relocation/ [Last accessed: 13th February 2025].

Climate ADAPT (2024). *Landslides*. European Climate and Health Observatory, Climate ADAPT. Available from: https://climate-adapt.eea.europa.eu/en/observatory/evidence/health-effects/landslides [Last accessed: 13th February 2025].

Fidan, S., Tanyaş, H., Akbaş, A., Lombardo, L., Petley, D. N., & Görüm, T. (2024). Understanding Fatal Landslides at Global Scales: a Summary of Topographic, Climatic, and Anthropogenic Perspectives. *Natural Hazards*, 120(7), 6437–6455. https://doi.org/10.1007/s11069-024-06487-3.

Froude, M. J., & Petley, D. N. (2018). Global Fatal Landslide Occurrence from 2004 to 2016. *Natural Hazards and Earth System Sciences*, 18(8), 2161–2181. https://doi.org/10.5194/nhess-18-2161-2018.

Ganepola, G. A. C., & Jayawardena, U. (2020). *Effect of temperature increase for the occurrence of natural landslides*. International Conference on Geotechnical

Engineering, Colombo. Available from: https://www.researchgate.net/publication/356926856_Effect_of_temperature_increase_for_the_occurrence_of_natural_landslides [Last accessed: 13th February 2025].

Gariano, S. L., & Guzzetti, F. (2016). Landslides in a Changing Climate. *Earth-Science Reviews*, 162, 227–252. https://doi.org/10.1016/j.earscirev.2016.08.011.

Guzzetti, F., Gariano, S. L., Peruccacci, S., Brunetti, M. T., & Melillo, M. (2022). Chapter 15 - Rainfall and Landslide Initiation. In: R. Morbidelli (Ed.) *Rainfall*. Elsevier, pp. 427–450. https://doi.org/10.1016/B978-0-12-822544-8.00012-3.

Ietto, F., Conforti, M., Tolomei, C., & Cianflone, G. (2022). Village Relocation as Solution of the Landslide Risk, Is It Always the Right Choice? The Case Study of Cavallerizzo Ghost Village (Calabria, Southern Italy). *International Journal of Disaster Risk Reduction*, 81, 103267. https://doi.org/10.1016/j.ijdrr.2022.103267.

IPCC (2021). *Climate Change 2021: The Physical Science Basis. Contribution of Working Group I*. Cambridge University Press.

ISPRA (2011). *World Landslides Forum: Each year, landslides cause damage to 6 billion Euros*. Available from: https://www.isprambiente.gov.it/contentfiles/00010100/10185-second-world-landslide-forum-press-release.pdf [Last accessed: 13th February 2025].

Jemec Auflič, M., Bezak, N., Šegina, E., Frantar, P., Gariano, S. L., Medved, A., & Peternel, T. (2023). Climate Change Increases the Number of Landslides at the Juncture of the Alpine, Pannonian and Mediterranean Regions. *Scientific Reports*, 13(1), 23085. https://doi.org/10.1038/s41598-023-50314-x.

Loche, M., & Scaringi, G. (2025). Assessing the Influence of Temperature on Slope Stability in a Temperate Climate: A Nationwide Spatial Probability Analysis in Italy. *Environmental Modelling & Software*, 183, 106217. https://doi.org/10.1016/j.envsoft.2024.106217.

Luo, H. Y., Zhang, L. M., Zhang, L. L., He, J., & Yin, K. S. (2023). Vulnerability of Buildings to Landslides: The State of the Art and Future Needs. *Earth-Science Reviews*, 238, 104329. https://doi.org/10.1016/j.earscirev.2023.104329.

Pollock, W., & Wartman, J. (2020). Human Vulnerability to Landslides. *Geohealth*, 4(10), e2020GH000287. https://doi.org/10.1029/2020gh000287.

Shaw, R., Takeuchi, Y., & Rouhban, B. (2009). Education, Capacity Building and Public Awareness for Disaster Reduction. In: K. Sassa & P. Canuti (Eds.) *Landslides – Disaster Risk Reduction*. Springer Berlin Heidelberg, pp. 499–515. https://doi.org/10.1007/978-3-540-69970-5_26.

Sim, K. B., Lee, M. L., & Wong, S. Y. (2022). A Review of Landslide Acceptable Risk and Tolerable Risk. *Geoenvironmental Disasters*, 9(1), 3. https://doi.org/10.1186/s40677-022-00205-6.

Stanley, T. A., Soobitsky, R. B., Amatya, P. M., & Kirschbaum, D. B. (2024). Landslide Hazard Is Projected to Increase Across High Mountain Asia. *Earth's Future*, 12(10), e2023EF004325. https://doi.org/10.1029/2023EF004325.

Tagarelli, V., & Cotecchia, F. (2020). The Effects of Slope Initialization on the Numerical Model Predictions of the Slope-Vegetation-Atmosphere Interaction. *Geosciences*, 10(2), 85. https://doi.org/10.3390/geosciences10020085.

Tian, N., Lan, H., Li, L., Peng, J., Fu, B., & Clague, J. J. (2025). Human Activities Are Intensifying the Spatial Variation of Landslides in the Yellow River Basin. *Science Bulletin*, 70(2), 263–272. https://doi.org/10.1016/j.scib.2024.07.007.

Tichavský, R., Ballesteros-Cánovas, J. A., Šilhán, K., Tolasz, R., & Stoffel, M. (2019). Dry Spells and Extreme Precipitation Are The Main Trigger of Landslides in Central Europe. *Scientific Reports*, 9(1), 14560. https://doi.org/10.1038/s41598-019-51148-2.

WHO (2025). *Landslides*. World Health Organization. Available from: https://www.who.int/health-topics/landslides#tab=tab_1 [Last accessed: 13th February 2025].

WMO & UNDRR (2022). *Early Warnings for All: The UN Global Early Warning Initiative for the Implementation of Climate Adaptation, Executive Action Plan 2023-2027*. United Nations Office for Disaster Risk Reduction (UNDRR); World Meteorological Organization (WMO). Available from: https://library.wmo.int/idurl/4/58209 [Last accessed: 13th February 2025].

World Bank (2020). *The Global Landslide Hazard Map: Final Project Report*. Available from: https://datacatalog.worldbank.org/search/dataset/0037584/Global-landslide-hazard-map [Last accessed: 13th February 2025].

Xu, Q., Zhao, K., Liu, F., Peng, D., & Chen, W. (2021). Effects of Land Use on Groundwater Recharge of a Loess Terrace Under Long-Term Irrigation. *Science of the Total Environment*, 751, 142340. https://doi.org/10.1016/j.scitotenv.2020.142340.

Index